Postcards *from the* Brain Museum

BRIAN BURRELL

Postcards

from the

Brain Museum

The Improbable Search for Meaning in the Matter of Famous Minds

BROADWAY BOOKS
NEW YORK

PRINTED IN THE UNITED STATES OF AMERICA

Visit our website at www.broadwaybooks.com

First edition published 2004

BOOK DESIGN BY JENNIFER ANN DADDIO

Library of Congress Cataloging-in-Publication Data
Burrell, Brian
Postcards from the brain museum : the improbable search for
meaning in the matter of famous minds / Brian Burrell.
p. cm.
Includes index.
1. Neuroanatomy. 2. Brain—Localization of functions.
3. Anatomical specimens. 4. Anatomical museums.
5. Gifted persons. 6. Criminal anthropology. I. Title.

QM451.B883 2004
153—dc22
2004045648

ISBN 0-385-50128-5

1 3 5 7 9 10 8 6 4 2

For

James Hathaway, teacher,

and

Dave Carson, carpenter,

who

taught

me

to

think

"Wake up and look facts in the face. Here we have a fiend whose brain—"

"Whose brain must be given time to develop. It's a perfectly good brain, Doctor. You ought to know. It came from your laboratory."

"The brain that was stolen from my laboratory was a criminal brain!"

(Henry stops, looks back at the creature's room, and shrugs.)

"Well, after all, it's only a piece of dead tissue."

HENRY FRANKENSTEIN TO DR. WALDMAN,
FROM UNIVERSAL STUDIOS' 1931
VERSION OF *FRANKENSTEIN*

Contents

Caveat

If you have never explored the recesses of an anatomical museum—brace yourself. It takes some getting used to. It is not so much the sight of it, the shock of the uncanny (as Freud described the contemplation of familiar objects in unnatural settings), as the smell of it. Not the smell of death, as you might expect, or even the reassuring whiff of decay (which at least promises an eventual end to things), but a smell that reneges on any hope of oblivion. It is formaldehyde, a fixative so powerful that it does to living cells what the pause button on the remote control does to pixels. And it does not discriminate. It will do the same thing to the skin of a careless anatomist that it does to the specimen he is preparing. And when you first walk into a closet full of preserved human brains, it will smell as though it wants to do the same thing to you.

Postcards *from the* Brain Museum

Introduction

IN THE SPRING of 1815, the volcanic peak of Tambora in the East Indies exploded, causing a tidal wave that killed over fifty thousand people. It was the largest volcanic eruption ever recorded, and it continued for months, expelling at least five times as much magma as Krakatoa, and thirty times as much ash as Mount Saint Helens. Its massive plume darkened skies around the world for more than a year, bringing dramatic climate changes as far away as Western Europe, including the heavy rains that were thought to have been the decisive factor at the battle of Waterloo.

The eruption of Tambora had other, unexpected consequences. It blotted out the sun for such an extended period that in the Northern Hemisphere, 1816 became "the year without a summer." Years later, the novelist Mary Shelley (at the time still Mary Godwin) would look back on the event as a turning point in her life. In June of that fateful year, she and her future husband, Percy Bysshe Shelley, were vacationing with their friend Lord Byron at his villa on Lake Geneva. The rain was so unrelenting that after a week trapped indoors, the group had almost run out of diversions.

"We will each write a ghost story," suggested Byron.[1]

It had started promisingly enough a few days earlier, with Byron and Shelley arguing over philosophy and natural science, even the principle of life itself. They discussed the experiments of the galvanists, followers of Luigi Galvani and Alessandro Volta, who electrified frogs legs and made them jump and twitch, or stimulated the severed nerves of decapitated heads fresh from the guillotine with similar uncanny results. Mary listened intently as another guest, Dr. John Polidori, recalled a curious experiment by Erasmus Darwin (Charles Darwin's grandfather) which seemed to suggest the possibility of imparting life to a wet noodle, actually a piece of vermicelli.[2] But as the ash plumes of Tambora rained down upon the Alps, the wellspring of conversational topics began to run dry.

After Byron made his suggestion, he went off to write a vampire story. Polidori concocted a tale about a skull-headed lady. Mary at first couldn't think of anything. Then she had a dream, a "waking nightmare," as she later would call it: "I saw the hideous phantasm of a man stretched out, and then, on the working of some powerful engine, show signs of life and stir with an uneasy, half-vital motion."[3]

SINCE 1818, WHEN *Frankenstein* was first published, there have been countless retellings of Mary Shelley's tale of a mad scientist who builds a man out of spare parts and brings him to life. In the course of those retellings the creature has changed to the point that there are now two monsters—Mary Shelley's and Boris Karloff's. One is an athletic superman who talks incessantly, while the other plods along and never speaks. One is driven to murder by his fragile ego, the other by a defective brain. One is wounded by careless nurture, the other by cruel nature. Shelley's monster imposes his will on a world that rejected him. His modern counterpart, on the other hand, has no will at all. He is simply a prisoner of his misshapen brain.

In the film, the scene that illuminates the difference is one of the most famous in all of cinema. The setting is a lecture hall at Goldstadt

Medical College. As the scene opens, a professor named Waldman is gesturing to a jar containing "one of the most perfect specimens of the human brain that has ever come to my attention at this university." The jar next to it, he says, holds the brain of a typical criminal.

"Observe, ladies and gentlemen, the scarcity of convolutions on the frontal lobe as compared to that of the normal brain, and the distinct degeneration of the middle frontal lobe." These features, Dr. Waldman points out, "check amazingly with the case history of the dead man before us, whose life was one of brutality, violence, and murder." He announces that he will leave the specimens on the demonstration table should anyone care to take a closer look. Class is dismissed.

As the room clears, no one shows the least bit of interest in the two brains, except for a hunchbacked man, who is seen entering through a window at the back of the lecture hall. His name is Fritz, and he is on a very important mission.

This scene, from Universal Studios' classic film, does not appear in the 1920s London stage play upon which it was based, nor does it appear in Mary Shelley's novel. It was inserted at the insistence of the film's twenty-one-year-old producer, Carl Laemmle Jr., in order to motivate the plot. Laemmle felt that neither the play nor the novel provided a clear enough reason for the creature to go berserk, so he asked the film's original director and screenwriter, Robert Florey, to invent something that would justify the eventual disaster. It was Florey who came up with the criminal brain.

As anyone who has seen the film knows, the scene has a darkly comic ending. While trying to make off with the healthy brain, Fritz is startled by a sudden noise and drops the jar. In its place, he snatches up the jar labeled DISFUNCTIO CEREBRI. The rest, as they say, is cinematic history. It was also, perhaps unintentionally, a clever send-up of a half century of neuroscientific discovery. As it turned out, Robert Florey may have done more to shape the popular perception of the human brain than anyone before or since.

FRANKENSTEIN THE MOVIE was made in 1930, the same year that the journal *Science* reported that a group of German and Soviet scientists had found an explanation for the peculiar genius of Vladimir Lenin in the third cell layer of his cerebral cortex. Over the previous half century, studies of this kind had been appearing in scientific journals with some regularity. Detailed monographs on the preserved brains of the historian George Grote, the blind-deaf-mute Laura Bridgman, the explorer John Wesley Powell, and many others had boldly proposed anatomical explanations for their intellectual accomplishments. At the same time, physical anthropologists, notably Cesare Lombroso in Italy, were conducting studies on the brains of executed criminals in the hope of isolating the causes of criminal behavior.

This was not science fiction, but cutting-edge anatomical research, and it was taken very seriously, so seriously that in the late 1800s, mutual dissection societies began to spring up in Europe and the United States, for the express purpose of accumulating brains of a "high type." The membership rolls included not only scientists, but artists, musicians, writers, politicians, and businessmen. By century's end, leaving one's brain to science had become so fashionable that such disparate personalities as Walt Whitman, William Makepeace Thackeray, Ivan Turgenev, and Hermann von Helmholtz were swept up in the cause. So too, involuntarily, were hundreds of executed criminals. In various laboratories around the word, collections of special brains in jars were such a common sight that by the time Robert Florey sat down to doctor the script of *Frankenstein,* he did not have to invent very much. Dr. Waldman's lecture on degenerate lobes, for example, was lifted from the literature of criminal anthropology. The juxtaposition of a normal and a criminal brain was already a cliché.

All of which points to a seemingly simple question: Is there really anything to it? Can anyone read the cast of someone's mind in the shape of his or her brain? Has neuroscience progressed to the point where the experts can distinguish the brain of a genius from that of a madman, for example, simply by examining the folds of the hemispheres or the structure of the cells? The not-so-simple answer to that question (a less than resounding "not yet"), and the strange path this quest has taken, is the subject of this book.

THE THOUGHT OF removing a human brain suggests an act of such extreme violence and brutality as to be obscene. Yet it happens every day as a practical matter. The specimens are useful for obvious reasons—to show students what a brain looks like at different stages of its development, to show what it looks like in different stages of disease. In other words, scalps are routinely parted, skulls caps split open, and their contents removed and preserved in order to find out something about how a person lived or died.

Hundreds of research laboratories around the world maintain brain banks, and each year acquire thousands of specimens from people who leave their bodies to science, some willingly and some unknowingly. In what has lately become a common occurrence, the British government recently announced that between 1970 and 1999 some twenty-two thousand brains had been removed during autopsies and sold off to brain banks without the permission of surviving relatives. (The going rate was about sixteen dollars per brain.)[4] The country's inspector of anatomy estimated that the practice had been epidemic since about 1960, and the British are not alone. Since the mid-1800s, when the techniques for preserving so-called "wet specimens" were perfected, the demand for human brains has far exceeded the supply. It is easy to see why.

Preserved brains are essential to research on damage caused by injuries or brain-related diseases—notably Huntington's chorea, Alzheimer's disease, brain tumors, or strokes. The Army Medical

Museum, now incorporated into the National Institutes of Health, has tens of thousands of such brains, with special emphasis on battlefield head wounds. (They also have a few famous specimens, including the brains of Abraham Lincoln and the assassins of James A. Garfield and William McKinley. John F. Kennedy's brain, rather famously, has gone missing.) McLean Hospital, a famous Massachusetts psychiatric clinic associated with Harvard University, maintains one of the largest brain banks in the world, providing research material for, among other things, studies of schizophrenia and manic depression.[5] Another collection of diseased brains, recently discovered in the basement of a dormitory at Yale University, contains hundreds of examples of the kinds of skull-penetrating tumors that are rarely seen today.[6] The collection belonged to the pioneering neurosurgeon Harvey Cushing, who used it to develop the surgical techniques that have made such horrifying afflictions a thing of the past.

If collections of this type seem gruesome, they are undeniably necessary. They have yielded up information that has resulted in the alleviation of untold suffering. But there are other repositories of brains that are dedicated to a very different proposition, and these were not so much necessary as inevitable. Rather than preserving brains in order to find out what went wrong with them, they were founded in the hope of discovering what went right.

FOR MUCH OF the nineteenth century, divining the quality of the mind from the shape of the head was the business of phrenologists. Because methods for preserving whole brains were not perfected until halfway through the century, practitioners of this reviled pseudoscience focused instead on bumps on the skull, in the belief that they accurately reflected the contours of the brain. They do not, as it turns out, and the idea's supporters would pay dearly for this mistake. Yet in some respects the phrenologists were actually ahead of their time.

The guiding principle of phrenology is that the brain has many parts, and that it divides the labor of thinking among them. Examine the shape of the brain and the size of these parts, the theory claims, and you will discover the mind's strengths and weaknesses. At first, this led scientists to collect skulls and plaster busts of famous men. But by the late 1800s, with the bump-readers' credibility in ruins, the emphasis shifted from the skull to the brain itself, to its size, shape, and fissural patterns. Phrenology then went forward under a new name—cerebral localization—and the call went out for brains instead of skulls.

The first brain donation society, dubbed by its founders with intentional irony as the Society of Mutual Autopsy, was established in Paris in 1876. Through an unusual fraternal bond, men of science, the arts, business, and politics, most of them freethinkers (essentially atheists and revolutionaries), banded together to help science unlock the mysteries of the mind by searching for the substrate of intelligence in the material, rather than spiritual, realm. Some readers may be familiar with the society, at least indirectly, through a famous essay—"Broca's Brain"—in which the astronomer Carl Sagan describes a behind-the-scenes tour of Paris's Musée de l'Homme. While rummaging through a moth-balled anthropological collection from the previous century, Sagan stumbled upon a cabinet full of preserved brains, which got him thinking about the Faustian bargains that scientists sometimes enter into, out of either curiosity or hubris. Coming upon the brain of Paul Broca, a pioneering anatomist who discovered the speech center of the cerebral cortex, Sagan was moved to say, "These forgotten jars and their grisly contents had been collected, at least partly, in a humanistic spirit; and perhaps, in some era of future advance in brain studies, they would prove useful once again."[7] This is true enough. The brains were indeed collected in a humanistic spirit, although the motivation behind the autopsy society was mostly political. The brain donors founded their society in order to advance the cause of science at the expense of organized religion. Which is to say, there was more to the story than met Sagan's discerning eye.

The subject of famous brains in jars turns up in another work of popular science, Stephen Jay Gould's 1979 bestseller, *The Mismeasure of Man*. Gould's subject is an aspect of biological determinism, the argument that biology is destiny, that physical attributes of the races determine their place in the social order. Specifically, Gould takes aim at the contention that individuals can be rated on a numerical scale of worth using intelligence testing. The first such tests, he points out, were biological, or, to use the technical term, *anthropometric*. By measuring attributes such as the volume of the skull, the weight of the brain, or the relative sizes of the lobes, some scientists (including Paul Broca) contended that individuals and racial groups could be ordered on an evolutionary scale of development. Gould devotes the first part of his book to debunking this theory, and to exposing the shoddy science behind it. But his real target, which takes up most of the book, is the invention of the intelligence quotient, or IQ, and the development of written tests that purport to measure it.

The political ramifications of intelligence testing are so extensive that the subject remains mired in controversy, and when Gould's book first appeared it was subjected to intense scrutiny. Some writers criticized him for (among other things) his portrayal of Paul Broca as a reckless experimentalist who was motivated by a racist agenda. Broca, it should be said, did hold racist beliefs, but the fact that they were not the sole, or even the primary, motivation for his researches went beyond the scope of Gould's book, and would have diluted his essential argument. Gould acknowledged that there is always a "broader historical context," and he purposely limited his critique to the research itself, leaving to others the task of fitting the "scientific claims into social settings." With respect to the anthropometric study of the human brain, this "fitting" is one of the goals of this book.

PROPERLY SPEAKING, THERE are no brain museums. There are museums that have famous brains but keep them hidden from sight. At the Institute of the Brain in Moscow, for example, the brains of Vladimir Lenin, Joseph Stalin, Vladimir Mayakovsky, Maxim Gorky, Sergei Eisenstein, and many others are kept under lock and key. Visitors and foreign researchers are kept out. The same is true of an even more impressive collection at the University of Tokyo, which contains some 120 specimens, including the brains of three prime ministers and scores of Japan's most revered writers and poets. Another collection, in Düsseldorf, Germany, is so extensive that it has not yet been catalogued, which prevents its curator from disclosing the names. From archival sources, it is known that the gray matter of many prominent scientists, including a few Nobel Prize winners, can be found there.

Fortunately there are exceptions to this rule of secrecy. The most accessible of these is not properly a brain museum as much as a display cabinet outside the offices of the psychology department at Cornell University, in which eight brains are on permanent exhibition. They belonged to men who achieved a degree of eminence or notoriety (seven were professors, one a mass murderer), and would be forgotten today if not for their presence in the cabinet. A similar collection in Göttingen, Germany, contains the brains of four professors who died in the 1850s, including that of Carl Friedrich Gauss, one of the three greatest mathematicians of all time (a ranking, by the way, that has nothing to do with the proportions of his brain). The Göttingen collection is the oldest collection of so-called "elite brains" in the world, and for that reason is the most important, but its survival was a happy accident. For over a century it was considered so unimportant that it was left undisturbed.

If there are no brain museums as such, there are many anatomical collections that have a famous brain or two on display. The most notable of these is the Hunterian Museum at London's Royal College of Surgeons, which possesses the brain of Charles Babbage, another mathematician, and the inventor of the Difference Engine, the nineteenth-century precursor to the digital computer. Babbage's is perhaps the most pristine brain specimen extant. Each of his bleached-white hemispheres, his own personal computers, floats in its own rectangular jar in preservative fluid as clear as spring water. The brain seems more than disembodied. In fact, it is difficult to associate it with *any* body, so clean and well maintained are its surroundings. But Babbage's brain is an exception to a rule of neglect that plagues most of these collections. Like goldfish, brains require routine maintenance, and when they don't get it, the results are not pretty.

In Paris and in Philadelphia there are modest brain collections that may be viewed by special permission, although their curators are naturally suspicious of anyone who asks for a look. Not many people do, which is fortunate, because most of the specimens have been left to their fate over the last hundred years. Readers of "Broca's Brain" should be forewarned that nothing quite so presentable as Babbage's brain awaits them at Paris's Musée de l'Homme. The remnants of the Society of Mutual Autopsy are still there, in storage, much as Sagan described them. (Broca's brain itself is not; it now resides across town at the Musée Dupuytren at the old Faculté de Médecine, where Broca worked.) What is surprising, if not disappointing, is something that Sagan seems not to have noticed: the deplorable state of the brains that inspired his reverie. In most of the jars the fluid is so cloudy that the specimens cannot be seen. Pieces of decayed tissue float in suspensions that look more like bouillabaisse than brains. In at least one jar, the fluid has evaporated altogether, leaving behind a few shards that could easily be mistaken for bits of charcoal.

BY ALL APPEARANCES this might seem to be a closed chapter in the history of science. Most of these brains were never studied;

those that were yielded few clues to the source of their owners' eminence or genius. But the fact that nothing has been proven (and that many things have been disproven) does not rule out the possibility that, as Sagan suggested, these brains might "prove useful once again." Today, at least a few researchers seem to think so.

In 1989, Russian scientists managed to acquire the brain of Andrei Sakharov, which they installed at their Institute of the Brain and carefully prepared for study. As of this writing, they are still studying it. In 1999, the British medical journal *The Lancet* published the results of a study that set out to make what the Russians have called a "post-mortem diagnosis of genius" on the brain of Albert Einstein. (It suggested that the peculiar structure of Einstein's parietal lobes might have contributed to his discoveries.) Still more recently, news sources revealed that the brains of three members of Germany's infamous terrorist group, the Baader-Meinhof gang, were removed without permission after their deaths in prison in the 1980s, and were studied by a neurologist who claims to have found "neurological abnormalities." It is not clear what this means, yet the game goes on, often with little regard for what has gone before.

What follows is intended to make up for this lack of hindsight. It is the story of an irresistible and surprisingly recent idea, one that claims (quite reasonably) that the vessel of a distinctive mind should itself be distinctive in a significant way. This is, as one might expect, a story of brain snatching, of brain dropping, of Faustian bargains, of Frankenstein scenarios, of careers, lives, and professional reputations sacrificed to misguided ideas. But it is also a story of discovery and progress, of the emergence of the scientific method in the biological sciences, of the birth of the idea of the brain as the organ of the mind, and of the invention of such academic fields as physical anthropology and criminology.

Although this book dips into many realms, from philosophy, politics, and religion, to medicine, psychiatry, psychology, anthropology, and even phrenology, it is not highly technical. The chapters proceed chronologically, to the extent that such a thing is possible in a tale that ranges over centuries and continents, that is propelled by scientific discoveries, political agendas, and social movements, and that involves a few famous, some infamous, and several forgotten characters of history. One of its lessons is that in contemplating an object as complicated as the human brain, an imaginative researcher can always unearth evidence for just about anything he or she sets out to prove. Thus have brain-anatomical arguments been made for the supremacy of whites over blacks, of men over women, and of "educated and orderly persons" over criminals. It should be pointed out, however, that equally compelling evidence has been found to support the opposite of each of these claims. In other words, nothing has been proven beyond even the faintest shadow of a doubt.

It is one of the most remarkable aspects of the study of the human brain that, despite astounding advances in the technology of biological science, the big questions that researchers ask today are not very different from the questions they were asking over a century ago: How does the physical brain give rise to the mind? What does the mind consist of? What are its components, its attributes, its faculties? But as the story in these pages unfolds, it might be worth reflecting on a more basic question that will inevitably determine how far this line of inquiry can go. It simply asks whether it is possible for an object, specifically a human brain, to observe itself objectively, and in doing so come to a genuine understanding of what it sees.

Chapter 1

The Most Complex Object in the Universe

AS NEUROSCIENTISTS NEVER tire of pointing out, the adult human brain is the single most complex object in the universe, and one of the least understood. On average, it weighs about three and a half pounds, most of which is fragile, malleable tissue—so fragile that the brain is the most difficult part of the body to access, remove, handle, and study. A Soviet neuroscientist once likened its consistency to the insides of a watermelon, but even that overstates its structural integrity. If placed on a table, a fresh brain will quickly surrender to gravity and collapse into a heap (more like gelatin than watermelon). Within eight hours, it will begin to decompose, the first part of a dead body to do so. There may indeed be nothing so complex in the universe, nor anything quite as delicate.

The familiar shape of the human brain is somewhat misleading. As a ubiquitous graphic symbol, its most prominent feature, the massive, fissured cerebrum, has come to symbolize the unlimited potential of human thought, if not the very means of man's dominion over the planet. Yet it also bears an unmistakable resemblance to a comical turban, and for most of recorded history it was treated that way.

Until the 1600s, anatomists drew the brain's tortuous surface as a mass of undifferentiated folds, which they likened in their randomness to the folds of the small intestine.[1] After puzzling over its purpose, they concluded that the folds were nothing more than an apparatus for the manufacture of phlegm, which the brain squeezed out through the sinuses, and for producing tears, which it squeezed out through the eyes. Only in the last 150 years have scientists come to appreciate what really goes on in those folds, and that their rapid evolution, seemingly accomplished over the last million years, is easily the most impressive achievement in Darwin's universe. In hindsight, the human brain is a triumph of adaptation, so impressive both in size and reputation that until recently it has succeeded in hiding what has in common with the brains of all mammals, which turns out to be quite a bit.

The principal parts of the mammalian brain are the brain stem, the cerebellum, and the forebrain. The stem houses the physical plant. It monitors and regulates unconscious physical processes such as breathing, blood flow, digestion, and glandular secretion. It consists of the medulla, an extension of the spinal cord, a nodule called the pons, and a short connector called the midbrain. The cerebellum, or little brain, lies behind this assembly, and it is aptly named. With its striated exterior and dual hemispheres (at least in primates), it hangs behind the cantilevered back porch of the forebrain like a wasp's nest. Although its role is still not completely understood, the cerebellum is believed to act as a kind of automatic pilot for fine muscle control. If recent studies are correct, it also plays a role in short-term memory, attention, impulse control, emotion, cognition, and future planning. Researchers suspect that it might be a kind of backup unit, an auxiliary brain. Its loss, while far from desirable, is not fatal. The rest of the brain seems to be able to compensate. The forebrain, on the other hand, is indispensable. It is what makes humans human, and, as a result, the search for the anatomical locus of genius, criminality, or insanity begins there.

Neurologists tend to be of two minds about the forebrain. Some see it as two complementary but sometimes competing hemispheres, an uneasy coalition of rationality and impulse. Others attribute the same inner struggle to a cold brain and a hot brain, the entire cerebrum being the source of cool calculation, and a set of nested organs called the limbic system giving rise to hot instincts and urges. The left brain–right brain dichotomy originated in the 1960s when neurosurgeons intervened in acute cases of epilepsy by severing the corpus callosum, the fiber bundle that allows the two hemispheres to communicate with each other. In most cases the seizures went away, leaving patients with a curious split personality. The notion of a hot brain and a cold brain is somewhat older, and reflects a belief that higher functions, specifically the intellect, are situated literally and figuratively above the lower functions. Just as the intellect is supposed to keep the passions in check, the massive cerebrum envelops the limbic organs—the thalamus, hypothalamus, hippocampus, and amygdala—and, on good days, dominates them. Some psychologists like to refer to the embedded limbic system as the reptile brain, a term they invented as a way to market themselves to Madison Avenue and Hollywood. The impulsive animal brain, they say, seeks dominance, safety, or sustenance, and it wants everything NOW—everything from an ice cream sundae to a sport utility vehicle, with little concern for practicality or consequence. Without the intervention of the cold, rational brain, the reptile brain can act quite unreasonably in getting what it craves.

Whether we really are of two minds in a literal sense is far from proven. Yet there is no denying that every mind undergoes a constant struggle between reason and emotion, between impulse and hesitation, between short-term strategies and long-term planning. The conscious brain struggles to rein in the unconscious, to calm nameless fears and anxieties. But if human beings, which is to say the human brain, hot or cold, can be characterized by one driving force, it would have to be curiosity, which has allowed it to explain just about everything in the universe, with two notable exceptions—the universe and itself.

What distinguishes a human brain from an animal brain, from an actual reptile brain? The size of the cerebrum, for one thing, and thus its surface area. But size, it turns out, isn't everything. The brain of an elephant, for example, is about four times as large as a man's, a blue whale's almost six times as large. Neither, of course, can match the forty-to-one body-to-brain-weight ratio in humans, but if ratios were all that mattered, the lowly field mouse, with a body-to-brain ratio of eight-to-one, would sit at the head of the class. Although the thinking part of the cerebrum, its outer shell, is four times thicker in humans than in rats, and four hundred times greater in surface area, the difference between men and mice is several orders of magnitude larger than dimensions alone can explain. It is not so much a matter of size, as of cerebral specialization. As the Alexandrian physician Erasistratus guessed in the fourth century B.C., the advantage lies in the folds, which are more developed in man than in any of the beasts.[2]

Are the folds in the brains of geniuses different from the folds of ordinary folk? The possibility has haunted investigators for a century and a half, and still has its supporters. Although not the only candidate for the anatomical substrate of genius, the folds are easily the front-runner because, contrary to the writings of the ancient anatomists, they are not entirely random. And where there is a pattern, there is assumed to be a meaning. In order to understand these patterns, you will not need a medical degree and a copy of *Gray's Anatomy*. Instead, a thumbnail sketch should suffice.

TO APPRECIATE THE rudimentary topology of the folds of the brain, place your right hand on the table and make a loose fist, relaxing the index finger so that the tip of the thumb rests inside the crux of the first knuckle. In other words, turn your hand into a talking clam, a puppet. Now shut the clam's mouth and imagine the hand enclosed in a mitten. What you are looking at, roughly speaking, is the left cerebral hemisphere of a primate brain. To turn it into a human brain, exchange for the mitten a boxing glove.

The surface of a real brain, of course, is riven by fissures, but these can easily be supplied with a pen. Begin by drawing a heavy line across the knuckles. In a real brain, this line is called the central sulcus (the Latin word for fissure); in older texts it is called the fissure of Rolando, after the eighteenth-century Italian anatomist who first described it. Just in front of this line is the precentral gyrus (gyri, also known as convolutions, are ridges of tissue that lie between fissures); just behind the precentral gyrus lies another ridge called the postcentral gyrus. The first of these contains the motor cortex, which in the left hemisphere controls movement on the right side of the body. The arrangement is inverted—the highest part of the gyrus, nearest the crown of the head, controls the foot, then comes the leg, and so on down to the lowest part of the gyrus, near the temple, which controls the hand and face. The postcentral gyrus registers sensation in a similar mapping. Taken together, these two convolutions, running over the crown of the head, form what is called the sensorimotor cortex.

Another important fissure, the lateral or Sylvian fissure, does not need to be drawn. It coincides with the gap between the thumb and the hand in the boxing glove model, and like that gap, it is very deep. A third useful line of reference, the occipital sulcus, should be added to the picture as a light line running across the very back of the hand, an inch or so above the wrist. Although there are a few other important fissures, these three—the central, Sylvian, and occipital—allow the hemisphere to be divided into its principal parts, the lobes.

The division of the hemispheres into lobes did not come about until the 1850s, and is generally credited to a French anatomist named Louis-Pierre Gratiolet.[5] It may come as a shock to discover that the lobes are not separate and independent units, that Gratiolet divided them somewhat arbitrarily and named them out of convenience by borrowing the words that describe the adjoining bones of the skull. There are four of them—the frontal, parietal, temporal, and occipital bones—delimited by the sutures, the skull's clearly visible expansion joints. Before Gratiolet came along, anatomists referred to the lobes of the brain using descriptors such as "posterior" and "anterior." But what Gratiolet noticed, after comparing hundreds of primate brains, was that although the pattern of folds in each brain is unique, there is a noticeable regularity to each one, especially in the deep fissures, much like the lines on the palm of the hand. The three primary fissures noted above, Gratiolet discovered, exist in the otherwise smooth brains of apes. They are also the first fissures to develop in the human fetus. It seemed only natural to use them as boundary lines for the four lobes. (A fifth lobe, tucked in the deep fold of the Sylvian fissure, completed the picture. Gratiolet called it the central lobe, although it is now known as the insula, and also, rather evocatively, as the island of Reil.)

The four principal lobes are easy to locate on the glove model of the brain. The frontal lobe lies above the Sylvian fissure and in front of the central sulcus. It is, essentially, the part of the boxing glove that encloses the fingers. Behind the central fissure lies the parietal lobe (that is, the back of the hand). The thumb of the glove, below the Sylvian fissure, represents the temporal lobe, and it extends as far back as the knuckle nearest the thumb's web. The base of the thumb, including its large pad of muscle, stands in for the occipital lobe. The dividing line is the occipital fissure. To picture the insula, imagine a handkerchief tucked into the fist, as in a magician's sleight of hand, and open up the thumb to reveal it.

Pierre Gratiolet did not believe that his division of the hemispheres into lobes had any functional significance. He preferred to think that the cortex worked more or less as a unit. But within ten years of his death in 1865, he would be proved wrong. During the 1870s, researchers in Germany and England produced detailed maps of cortical functions by stimulating the exposed brains of dogs, cats, and monkeys with electrical shocks. Through painstaking (if not pains-giving) labors they succeeded in correlating regions to reactions, and in the process set off a worldwide humane movement of antivivisectionists. At the same time, and less controversially, other researchers attempted to match clinical observations of neurological symptoms with postmortem examinations of brain lesions and tumors (a field known as morbid anatomy). In 1861, for example, Gratiolet's friend Paul Broca, a pioneering surgeon, demonstrated two instances of damage to the third frontal convolution in the left hemisphere in patients suffering from aphasia, a condition marked by the inability to produce speech. A decade later, the German psychiatrist Carl Wernicke isolated several other cases in which damage to the temporal lobe destroyed the ability to comprehend spoken or written language, while speech production remained unimpaired.

A quick consultation of the fist model of the brain shows a correspondence between the proximal phalanx of the index finger and the third frontal convolution of the cortex. This is Broca's area, a left brain module (in right-handed people) responsible for the mechanical production of speech. Below it, in the web of the thumb (that is, in the temporal lobe), is Wernicke's area, which handles language processing and speech recognition.

Through similar investigations the visual cortex was found to be situated in the occipital lobe, and the acoustic areas in the temporal lobes, but the experiments went only so far. There remained (and still remain) areas in each lobe, especially in the frontal and parietal lobes, that are not assigned motor or sensory roles. These have been lumped together as the so-called association areas, and are believed to be connected with memory, spatial and temporal orientation, and the intellect.

Keep in mind that your actual hand, the right fist you are looking at, is controlled by a relatively large part of the motor cortex in the left hemisphere it represents, a fact noted even by some of the ancients. This contralateralism—left controlling right, and vice versa—arises in the spinal cord at a splitting and crossing just below the brain stem in a place known, rather exotically, as the decussation of the pyramids.

This is where the brain ends. Or at least it is where the pathologist cuts the spinal cord when removing it.

GIVEN THE COMPLEXITY of what lies beneath, the outward aspect of the cerebral hemispheres, especially in a preserved brain, is deceiving. Its bloodless, somewhat bloated, aspect suggests a solid, homogeneous core—a cauliflower, perhaps. The business of thinking, it might seem, goes on somewhere in the middle. But as it turns out, the real action takes place near the surface. What goes on *inside* is a good deal of networking.

If you take a wide blade and cut straight through a brain, the resulting cross section will reveal a thin beige rind bordering vast estuaries of pink. Once fixed in formaldehyde, the picture becomes clearer. The rind turns gray, giving rise to the popular term "gray matter," and the pink fades to white, becoming "white matter." The gray matter is the thinking apparatus. Only four millimeters thick in most places, it consists of nerve cells, or neurons, and support cells known as glial cells, in numbers so staggering that they can only be roughly estimated, like the grains of sand on a beach. In its entirety, the gray matter is usually referred to as the cerebral cortex or isocortex—some 30 billion (give or take a few hundred million) neurons worth of computational hardware that accounts for almost half of the neurons in the entire brain. (Surprisingly, the cerebellum, which has only a third of the surface area of the cerebrum, contains even more neurons in its own gray matter.)

Under magnification, the brains of rats, cats, and apes are difficult to distinguish from those of humans. Their cortical material is the same. The real difference is partly one of size, but mostly one of connectivity. As a brain grows, not simply by genetic instructions but in response to its environment, it develops thousands of virtual connections between neighboring neurons. These connections, called synapses, serve as pathways for electrochemical charges. Each neuron can support anywhere from a few thousand up to a hundred thousand synapses. In all, there are a staggering 100 trillion of them in the adult brain. (In computer terms, that would be 100 million megabytes.)

The white matter, it turns out, is in the business of communicating all of this activity between parts of the cortex, or between the cortex and the body. It is made up of millions of long, threadlike tails called axons that originate in the neuron bodies. Some pass through the corpus callosum and allow the left hemisphere to communicate with the right. Others run down the spinal cord, and can extend up to a yard in length. Each axon is sheathed in an insulating substance called myelin, which is what makes the white matter white. The entire network has been compared to the wires of a telephone exchange, which is an understatement. It is estimated that if all of the axons in a single brain were stretched out end to end, they would extend three times the distance from the earth to the moon.

In most places the thin rind of gray matter is made up of six microscopically distinguishable cell layers. In the womb, these take shape in reverse order, the innermost layer (number six) forming first, and the cells of each successive layer migrating through their predecessors to precise spots determined by the DNA, with the outer layer (number one) forming last. According to a long-held belief, the division of neurons is complete at birth: you have then all that you will ever have. The subsequent growth of the brain, from an infant's 350 grams to an adult's 1,500 grams, is due to the expansion of the neuron bodies, to the formation of additional glial cells, and mostly to the formation of the fatty myelin sheaths on the axon tails. Recent evidence has suggested, however, that neurons can and do form even in adulthood, although it is a small consolation compared to the die-off that begins in earnest after age twenty, claiming about fifty thousand neurons per day.

In its formative stage, as the brain expands to full size (which it reaches at about age seven), the cortex contrives to maximize its surface area within the more slowly expanding skull by folding upon itself, creating its characteristic pattern of ridges and fissures. The end result suggests shifting sedimentary layers piling into one another, throwing up mountain ranges, and creating fault lines. Below the surface, the cortex varies in overall thickness from area to area, and in the individual thicknesses and cellular compositions of its layers from place to place, suggesting complex processes like those that shaped the earth's mantle. The surface of each hemisphere, mapped out by variations in cell structure, seems to consist of about fifty clearly differentiated zones. In some cases the transitions from zone to zone coincide with surface fissures; in others, transitions as sharp as fault lines have no relation to surface morphology. As a result, the cerebral cortex presents two distinct maps, one a surface map, the other subterranean, that have as much in common as a bus map and a subway map. Each divides the thinking part of the brain into a different set of neighborhoods, resulting in a multiplicity of names for the same landmarks, a historical legacy that drives anatomy students crazy.

GIVEN SUCH A wealth of structural detail in what seems, at least at first glance, to be an undifferentiated mass of tissue, it seems perfectly natural to ask whether it is possible to see what is going on in any of those neighborhoods. In other words, was Dr. Waldman right? Can anyone read the shape of an individual brain?

The answer is: not yet. While no reliable anatomical markers for genius, criminality, insanity, epilepsy, dyslexia, or special talents have yet been found, many studies have suggested that such markers may exist. In theory they do exist. But proving it has not been easy.

In an age of dazzling technologies, it may seem that everything that there is to know about the structure of the brain is now known. But this is not the case. The brain maps described above are still highly subjective, and the location of brain functions is still a hotly debated topic. For example, fMRI and PET scans do show that Broca's and Wernicke's areas are involved in some way in processing and producing speech. But what are the structures involved? Where are their boundaries? What other areas of the cortex are involved?

It turns out that the compartmentalized view of the brain, the phrenological model in which clearly delimited zones play very specific roles in the thought process, greatly oversimplifies the organ and what it does. More and more, the brain is starting to look like a highly integrated device that contains backup systems (notably the cerebellum), as well as the capacity to reconfigure itself in order to compensate for injuries. This plasticity, as it is called, creates yet another problem for brain researchers. It implies that any given brain, like the mind it houses, is constantly changing.

So what do brain scientists look for in the preserved brain of a famous man or woman, genius or criminal? What can they hope to find? Because only a few scientists are still doing this, it might be more useful to ask what scientists looked for and found in the past, specifically in the recent past, because it was only two centuries ago that anyone thought to ask whether the shape of a brain held the key to a mind. Which is to say that prior to the 1800s, few people suspected that the brain's imposing exterior meant anything at all.

Certainly René Descartes didn't, although the great French philosopher provides a useful jumping-off point for this story, because it was Descartes who first suggested, quite inadvertently, that the human body, and thus the human brain, is really nothing more than a highly sophisticated machine.

Chapter 2

Descartes

THE SKULL OF René Descartes, the father of modern philosophy and the inventor of geometry's Cartesian coordinate system, may or may not reside in Paris's Musée de l'Homme. There is a skull in the collection that has his name on it, literally written between the sutures, but its provenance is in dispute. The reputation of its former owner, on the other hand, is not. According to legend, when the prominent phrenologist Johann Caspar Spurzheim noted the small size and diminished forehead of this skull, he instantly proclaimed, "Descartes was not so great a thinker as he was held to be!"[1] What else could he say? According to the prevailing wisdom of the day, great intellects were housed in large brains, and Spurzheim was not inclined to argue against evidence. But when some prominent French anthropologists took up the question a few years later, they doubted that a skull of such "mediocre dimension" could have belonged to the author of *Discourse on Method*.[2] Either the skull, the man, or the brain-size theory was a fake, and Spurzheim's learned colleagues knew better than to give up on the man. Had they been true Cartesians, they would have given up on the theory.

The specimen that inspired this soul-searching probably is a fake, but this would hardly have mattered to Descartes himself. During his lifetime, he established a reputation as a brilliant thinker, and a peripatetic one. He moved constantly, never settling in one place for very long. In 1649, Descartes reluctantly accepted a summons from Queen Christina of Sweden to become her private tutor, and within a year he would succumb to the cold climate and die of pneumonia. He was buried in Stockholm, but, true to his character, he did not remain there for very long. Seventeen years later his body was exhumed and taken to Paris. In the process, his head was cut off so that the body would fit into a special copper-clad coffin. At his reburial, at least two parts of his anatomy went missing—the skull, allegedly stolen by an army captain and sold to a collector, and an index finger, since lost. A skull later bought at auction by the Swedish chemist Johann Jakob Berzélius and donated to the French government is the one that now graces the museum's collection. It has a sufficiently rich history to make it an object of interest, and is usually exhibited as the real thing, which would not have bothered Descartes very much. As far as he was concerned, one skull was just as good as any other. He would have said the same thing about his brain.

When Descartes is remembered today, it is usually for one of three things: his invention of analytic geometry (in which geometric figures are represented as equations), his epistemology (a theory of knowledge derived from his most familiar formulation, "I think, therefore I am"), and a theory of mind and brain centered on a small gland named for the pine nut. Of the three, only his geometry has aged gracefully, and is still taught in the high schools. His philosophy, while still revered, is routinely stripped of all context and reduced to an empty motto. But it is his theory of the pineal gland that has fared the worst, inviting the kind of condescension and derision (how could anyone have believed such a thing?) that misses the point. Seen in its proper context, Descartes's awkward physiology of the soul was a significant breakthrough. It provided the crucial step that rescued anatomical science from its dark ages, and opened up the scientific investigation of the brain as the organ of the mind.

FROM ANCIENT TIMES down to the Renaissance there was little consensus about what purpose, if any, was served by the human brain. The Egyptians gave it almost nothing to do, and placed memory, the intellect, passions, wit, spirit, and all other mental processes in the heart. When preparing a body for burial, their embalmers would scramble the brains, draw them out through the nostrils with a hooked needle, and discard them. The heart, on the other hand, they carefully preserved in a special canopic jar that was interred with the mummy.

While the heart might have reigned supreme in Egypt, other organs had their admirers. The ancient Babylonians assigned emotion and spirit to the liver. The Mesopotamians hedged a bit, and placed emotion in the liver and intellect in the heart. But the heart's most powerful lobbyist was Aristotle, who famously contended that the brain serves as little more than a radiator, a cooling device for blood that is heated up in the heart, where the real business of thinking and feeling takes place. Not surprisingly, Aristotle got many of the details wrong, such as mistaking nerves for tendons, but his cardiocentric theory survived through hundreds of generations because few men bothered to check it against observable facts.

Among those who did check were some of the Hippocratic writers, who countered that "from nothing else but the brain come joys, delights, laughter and sports, and sorrows, griefs, despondency, and lamentations. And by this, in an especial manner, we acquire wisdom and knowledge, and see and hear, and know what are foul and what are fair."[3] In the second century A.D., Galen, the Roman anatomist and physician to the gladiators, also situated the mind (or soul) in the head. But in a decision dictated largely by dissection techniques, he placed it in the brain's cavities rather than in its solid parts.

Throughout antiquity (and even to the present day to some extent), what anatomists tended to see in the brain depended very literally on how they sliced it. Until the Renaissance, the preferred technique was to leave the brain in place while dissecting it. The approach was top down: first remove the skull cap, then make two or three thick horizontal slices, and observe the cross sections. This technique, which is dramatically illustrated in a famous series of woodcuts by the celebrated Renaissance anatomist Andreas Vesalius, reveals an impressive set of cavities called the ventricles. There are three of them (actually four, although the largest ventricles, numbered one and two, form a matched pair, and are treated as a unit). As the slicing progresses, the paired lateral ventricles are the first to appear. Seen in cross section, they resemble two symmetric French curves at the center of the brain. The next two slices reveal the much smaller third and fourth ventricles, which lie deeper and farther back in the brain stem. Because they are the most striking features that appear during the process, it was assumed that something important went on there.

In the fourth century A.D., Nemesius, a bishop from Emesa (in modern-day Syria) decided that this triune structure was no accident, and he placed sensation in the paired ventricles, imagination in the third ventricle, and memory in the fourth. The order was anything but random. In Nemesius's theory of mind, sensation comes first. All stimuli arrive in the brain at a central processor known as the sensorium commune, literally the "common sense," which sends them on to be evaluated and perhaps acted on in the third ventricle. In the final step they are stored as memories in the fourth ventricle.

The doctrine of ventricular localization, as Nemesius's theory came to be called, reigned supreme for over a thousand years in one form or another, and it was still the prevailing theory of mind when Descartes took up the study of the subject in the 1630s. Descartes would have been justified in discarding it, because it was based entirely on convenience. Instead, he set about finding a mechanism by which the ventricles could give rise to thinking and bodily movement. It never occurred to him that for all that Nemesius knew about physiology, he might just as well have placed the mental faculties in the ventricles of the heart. This partly explains why the ventricular theory could have persisted, in one form or another, for so long. It needed little evidence to sustain it and encountered nothing to refute it. It simply filled a void, or, more precisely, it filled three of them.

IF DESCARTES HAD been faithful to his own scientific method, he would have set aside the ventricular theory, along with the anatomical theories of Galen and Aristotle, and started from scratch. That was the whole point of his philosophical system, his so-called Method of Doubt: throw out every received idea, and rederive everything. For Descartes, this meant setting aside even his belief in God and his own soul. Instead of assuming their existence, he famously launched his career by setting out to prove it.

How Descartes arrived at his proof reads like a modern fable. In 1619, while serving with the Duke of Bavaria's army, Descartes was journeying home from the emperor's coronation when he took refuge in a small cabin. "I remained the whole day shut up alone in a stove-heated room," he later recalled, "where I had complete leisure to occupy myself with my own thoughts."[4] Few people have managed to spin quite as much out of a few spare hours. Descartes claimed that he passed the afternoon searching for "the means of gradually increasing my knowledge and of little by little raising it to the highest point." After a full day of such brain work, he drifted off into an uneasy sleep.

That night, Descartes had a dream in which he was visited by three visions. In the first, he saw himself in a windswept street in the company of friends. Because one of his legs was weak, he alone among the group struggled to remain standing. In the second vision, he was startled by a massive thunderclap, and he awakened to see his room inundated with sparks. In the third, a book of verses was presented to him by an old man, and, turning to the first page, he read the words: "What way of life shall I follow?"

Descartes interpreted the first dream as a reminder to be skeptical of the received wisdom that his friends adopted unquestioningly. The second dream was meant to prepare him for the blinding light of truth. The third showed him that there was indeed a path that would lead him to true knowledge. From that point on, he would "accept nothing as true which I did not clearly recognize to be so." The entire edifice of his thought, he decided, would thenceforth be built upon a committed skepticism reinforced by his three dreams. This was the buildup to his most famous observation.

> I noticed that whilst I thus wished to think all things false, it was absolutely essential that the "I" who thought this should be some thing, and remarking that this truth, *"I think, therefore I am,"* was so certain . . . I came to the conclusion that I could take it as the first principle of the Philosophy for which I was seeking.

In a short sequence of steps, Descartes would deduce the existence of God, then of his own soul, by reasoning that the fact of his thinking was entirely independent of his body. His brain, skull, skeleton, muscles, and organs were nothing more than the components of a sophisticated machine which served as a temporary home for his immortal spirit. Animals, he said, were all sophisticated machines. What distinguished men from the beasts, and confirmed their special place in the divine plan, was the added element of the soul.

Any attempt to reconcile the mind and the brain always comes down to a question that divides philosophers, psychologists, and neuroscientists into two opposing schools of thought. One is dualism, the idea that the mind has a separate existence from the brain. The other is called monism, and it says that the mind *is* the brain, or the product of physical processes that occur there. After his night in the cabin, Descartes staked out the first position, and it has been known as Cartesian dualism ever since. But he did not stop there. Descartes not only convinced himself of the body's and spirit's dual existence but he also set out to find the place where they joined together.

It was St. Augustine who laid down the official doctrine when he declared that the soul permeates the entire body. Ever since, Church scholars have been forced to explain how the process works. For example, what happens to the portion of the soul that inhabits an amputated limb? When does the soul first enter the body, and where? Descartes assumed that it had to have a point of access, probably in the brain, and very likely in the ventricles. Although he had little training in anatomy and was not very good at dissecting, he knew what he was looking for—something central, something in one of the ventricles. What he found was a small gland at the base of the third ventricle—the pineal gland—which seemed to be the only feature of the forebrain that was not doubled. Given the unitary nature of the soul and the central location of the gland, Descartes guessed that not only did this nodule serve as the soul's access point, but it also controlled the flow of vital fluids in the nervous tubes in such a way as to direct the body's motions. The pineal gland, he decided, was not only the body's link to the spiritual world, but its command and control center.

Unfortunately, Descartes's elaborate hydraulic model of motor function says more about his powers of imagination than observation. Its awkward linking of causes to effects is less suited to the elegantly articulated motions of a human body than to a Rube Goldberg invention, which is what Descartes's diagrams and descriptions suggest. For example, in his *Treatise on Man,* Descartes gives a detailed description of the nervous system. The text accompanying a picture of a boy with his foot near a fire reads: "if the fire A is near the foot B, the particles of this fire . . . have the power to move the area of the skin of this foot that they touch; and in this way drawing the little thread, c, that you see to be attached there, at the same instant they open the entrance of the pore, d, at which this little thread terminates, just as by pulling one end of a cord at the same time one causes the bell to sound that hangs at the other end."[5]

What he means is that external stimuli cause fluxes in the animal spirits, which travel to the brain via nervous tubes that terminate in the ventricles. The pineal gland registers the fluxes as an image, and it reacts to them (at the soul's discretion) by moving ever so slightly, opening or closing off tubes that allow spirits to flow back out to selected muscles, causing them to contract. Descartes compared the process to the elaborate system of fountains in the king's gardens, where a visitor's footstep could set off a series of allegorical displays. The "reasoning soul" manipulates the pineal gland, he said, much as the king's fountaineer operates the hydraulic machines in the royal grotto, "causing those machines to play several instruments or to pronounce several words."

But if the pineal gland runs the show, like the king's fountaineer, then what runs the pineal gland? The answer, of course, is no one. Or, as one of Descartes's harshest critics, the twentieth-century philosopher Gilbert Ryle, said (by way of assuring himself a permanent place in the quotations dictionaries), the driving force is a "ghost in the machine."[6]

Descartes's elaborate physiology of the nervous system won few converts in his day. As an anatomist, he is remembered less for what he accomplished than what he attempted—namely, a natural history of the brain and the mind. Fortunately, his shortcomings as a dissector did not detract from his reputation as a brilliant philosopher and mathematician. Nor did his anatomical mistakes diminish his importance in the history of the neurosciences. As preposterous as his theory of the soul may now sound, it took the crucial first step of explaining in mechanical terms not just the body, but the brain itself.

A DECADE AFTER his death in 1650, Descartes's *Treatise on Man,* a book he had written in the 1630s and then suppressed following the condemnation of Galileo, appeared in a French edition. Among its most enthusiastic readers was a young Danish anatomist named Niels Stensen, who is better known by his Latinized name—Nicolaus Steno. Steno is one of the scientific world's most unjustly forgotten men. When he is remembered at all today, it is for his solution to a long-standing puzzle that had baffled natural scientists for centuries. It had nothing to do with brains, and everything to do with rocks.

Since ancient times, travelers in Europe had marveled at the presence of seashells and fossilized marine life embedded in mountain rock. The most surprising of these finds were smooth triangular rocks called tongue stones. If it had not been for the altitudes at which they were found, tongue stones would have been taken for sharks' teeth, which is precisely what they looked like. But it made no sense. How could a shark's tooth become part of a mountain rock? Pliny the Elder claimed that the stones fell from the sky or the moon. Others believed that they actually grew in the rock. Steno, of course, had his doubts.

Early in his career, Steno established a reputation as an astonishingly skilled anatomist. His ability to tease apart tissue samples and distinguish subtle structures that eluded the gaze of others led to several important discoveries, including the salivary glands and tear ducts (a finding that undermined the humoral theory that tears were literally squeezed out of the folds of the brain). So when a great white shark washed ashore near Livorno in 1666, the fishermen brought the head to Steno, who made a careful dissection. After comparing the shark's teeth to the tongue stones, he concluded that they were all the same. He was not the first person to make this claim, but he was the first to propose an explanation, one that led him beyond the realm of anatomy into geology. The mountains, he said, had once been at the bottom of the sea.

Steno's great contribution to geology grew out of this discovery. According to Steno, the earth's crust originally formed in successive layers, with newer layers settling on top of old. Subsequent disruptions of this pattern, he added, were caused by volcanic disturbances and upheavals. This brilliant deduction resulted in geology's one bedrock principle—the law of superposition—which is also known as Steno's law.[7]

As an anatomist, Steno contributed a mere footnote to the history of the subject, but it is an interesting footnote. When he came upon Descartes's *Treatise* in 1660, Steno was immediately struck by its description of the heart. Descartes, taking a page from Aristotle, had argued that the heart acts as a furnace, that it forces blood through its chambers by heating it. Having dissected many hearts, Steno knew that this was wrong, that the heart is a muscle, and that it pumps blood by contracting. If Descartes could be so wrong on such an important question, Steno mused, how reliable was the rest of his anatomical work? And what about his theory of the brain?

Steno knew something about brains. More accurately, he knew just how much he didn't know. In 1662, he conducted a series of brain dissections that served as the basis for a series of lectures he delivered the same year in Paris. He approached the subject using the Cartesian method—which is to say, he discarded everything he'd ever read on the subject and assumed nothing. "Instead of promising to satisfy your inquiring minds about the anatomy of the brain," he informed his audience, "I confess to you here, honestly and frankly, that I know nothing about it."[8] The available textbooks, he decided, were useless, their authors completely unreliable. In a justly famous passage, he laid down what could still serve as the bedrock principle of neuroanatomy. "Every anatomist who dissects the brain," he wrote, "demonstrates from experience what he advances. The soft and pliable substance so readily yields to every motion of his hand, that the parts are imperceptibly formed in the same manner as he had conceived them before dissection."[9]

Steno's onetime hero Descartes was no exception. What Steno discovered when he looked at the pineal gland was a structure that was fixed firmly in place by delicate tissues. It could not move in the way

Descartes had described. There were no surrounding tubes. In looking back on the episode years later, he would say, "I do not reproach Descartes for his method, but for ignoring it."[10]

IN HIS LECTURES on the brain, Nicolaus Steno delivered what should have been a death blow to Descartes's clever physiology of the soul, but Steno himself ran into a conflict of interest before he could propose anything better. In 1667, after witnessing a Corpus Christi procession in Livorno, he was swept up by the spectacle and soon converted to Catholicism. "Either this host is no more than a piece of bread, and they are fools who pay it such homage," he wrote, "or it really does contain the body of Jesus Christ, and why do I not also honor it?"[11] Eight years later he took holy vows as a priest, and before long he was elevated to bishop. He was then sent to minister to a small Catholic congregation in Germany, which was tantamount to banishment for life. Steno, the most talented experimental scientist of his generation, gave up science for a higher calling, and the thread of his anatomical investigations was left to be taken up by others.

Fortunately, someone was waiting to pick up the thread. His name was Thomas Willis, and he was England's leading physician and a professor at Oxford College. In 1664, Willis published *Cerebri anatome,* the most accurate study of the human brain to date, with illustrations by the architect Christopher Wren. It was a dramatic improvement over Vesalius's celebrated text (with woodcuts by the studio of Rafael), and the anatomical facts it contained spelled doom for the Cartesian theory. Willis confirmed Steno's finding that the pineal gland could not act in the way Descartes had proposed, but he went even further. In a breakthrough made possible by improvements in dissection techniques, Willis dismantled the theory of ventricular localization that had been in place for over a thousand years.

Willis was one of the first anatomists to remove the whole brain from the skull and dissect it from the bottom up. It is unlikely that he would have made his discoveries without using this innovative technique. With recourse only to the most primitive methods for stabilizing the specimen, the task was a delicate one, but Willis noticed that as soon as the brain was removed the ventricles collapsed, which convinced him that they were nothing more than "complications of the brain's infolding."[12] As he sliced his way up from the brain stem, he could see fiber tracts splitting off from the medulla and migrating to the brain's outer rind, and not to the ventricles. By following the network of fiber bundles connecting the brain to the body, Willis could also see that the origination point of the whole system was the cerebral cortex. There was only one possible explanation: Thinking did not take place in the hollows of the brain, as the ancients had assumed, but in its solid parts.

When it came to filling in the details, Willis fell back upon the basic plan of localization that was already in place. He reassigned the mental faculties, placing memory in the gray matter and imagination in the white matter. All involuntary movements he attributed to the cerebellum. Descartes may have started the process, but it was Willis, with his superior dissection technique, who got the credit for inventing modern neurophysiology. By taking Steno's work to its logical conclusion, he provided the first real candidate for a paradigm of brain function.

"A PARADIGM IS a scientific achievement that for a time supplies the foundation for further scientific practice in a field of inquiry."[13] So wrote the science historian Thomas Kuhn in his famous *Structure of Scientific Revolutions* (1962), in which he showed how scientific knowledge progresses not necessarily in a straight line to the truth, but from one paradigm to the next.

A good example of a paradigm is the Ptolemaic belief that the earth is the center of the solar system, an idea that was motivated by religious belief. For centuries astronomers struggled to reconcile it with their observations of the heavens, but it was difficult work. The paradigm seemed a bad fit. With the invention of the telescope, the accumulation of irreconcilable facts reached a critical mass, and a new paradigm, championed by Copernicus and Galileo, began to win converts. Galileo was forced to recant, but a new generation of thinkers, including Descartes, saw through the charade, and the rout was on. As more scholars abandoned the Ptolemaic theory for the Copernican one, a "paradigm shift" ensued.

According to Kuhn, the essential criteria for a scientific paradigm are simple: it must be sufficiently unprecedented to attract a group of disciples, and it must be open ended enough to give them something to do, some details to work out. Few paradigms come into the world fully formed. There are always mistakes and miscues that have to be smoothed around the edges, a painstaking and drawn-out process that Kuhn refers to as "mopping up."

Kuhn also acknowledges that many scientists struggle with revolutionary ideas that fall short of these criteria. These include crackpot theories, random good guesses, or ideas that are simply ahead of their time and fail to attract a following. He calls such ideas "candidates for a paradigm."

Kuhn's *Structure* was not a prescription for how science *should* be done, but for how it *had been* done throughout recorded history, and he drew most of his examples from the hard sciences that provided the best illustrations—physics, astronomy, chemistry. Biology, on the other hand, did not fit the scheme very well because it did not emerge as a distinct entity until the 1800s, and spent most of that century in a turf war with theology. As a result, until the year 1800, no theory of mind and brain had enough science behind it to qualify even as a candidate for a paradigm, and the investigation lingered in what Kuhn referred to as a "pre-paradigm stage."

Kuhn believed that in the pre-paradigm stage of any science, investigators proceed somewhat randomly. They indulge in "early fact gathering," a long period of observation, collection, measurement, and reflection, in which all facts seem equally relevant. "In the early stages of the development of any science," Kuhn notes, "different men confronting the same range of phenomena . . . describe and interpret them in different ways."[14] This situation changes dramatically, he adds, when a candidate for a paradigm emerges from the pack and gains a foothold, and at the dawn of the eighteenth century, this is precisely what was happening in brain anatomy.

Descartes started the process with his mechanical theory of mind, Willis then supplied the crucial second step by attributing mental processes to the solid part of the brain, but the third step would be the difficult one. Someone had to be bold enough to write the soul out of the equation, to say that the brain was nothing more than a thinking machine, and that thought was merely a mechanical process. Whoever dared to assert this publicly had to know that he was setting himself up for a fall.

IN 1745, A French physician by the name of Julien de La Mettrie wrote a book entitled *The Natural History of the Soul*, which so scandalized Paris that its author had to flee; in his absence the book suffered a fate that awaited him if he ever returned. It was burned in public by the hangman. Two years later, while living in Holland, La Mettrie threw all caution to the wind when he published the brazenly titled *Man-Machine*, and had to flee again. Eventually, he was taken in by Frederick the Great, who seems to have found him amusing, and La Mettrie settled down to a lifestyle that confirmed what everyone thought would come of his godless theories. He became a hedonist.

La Mettrie did not merely suggest that man was a machine. He supported his claim with evidence that was available to anyone who cared to look—the fact that certain animals continue to move about after having their heads cut off, for example. To La Mettrie, this suggested the existence of a vital force that permeates all parts of the body, not just the brain. He called it the motor principle. "Here are far more facts than needed," he wrote, "to prove in an incontestable manner that each small fiber or part of any organized body moves itself by a principle unique to itself."[15] La Mettrie further proposed that a higher-level force, a "more marvelous one," exists in the brain, and that it animates all of the lesser forces. The truly shocking part of his schema, however, was that he ended the causal chain there, and effectively eliminated the ghost from the machine. ("By this," La Mettrie concluded, "all that can be explained is explained.")

Following La Mettrie's lead, a generation of atheists stepped out of the shadows and began to publish similar theories in which the immaterial soul was replaced by a material mind. Without a soul to account for the phenomenon of free will, one question emerged as paramount, and it still remains to be settled. Does the human mind come into the world already equipped with a set of skills and tendencies, or is everything that it will become acquired through learning and experience? In terms that will sound more familiar to a modern audience, Which matters more, nature or nurture?

John Locke, who as a student at Oxford had attended Thomas Willis's lectures and dissections, made the case for nurture. It was Locke who first proposed the idea of the blank slate (or tabula rasa) as a model for the mind at birth. Locke and his followers (taking Aristotle's lead) argued that all knowledge, all understanding, everything that constitutes mind, originates in the senses. Locke did believe in the immortal soul, in the existence of innate faculties, but he assumed that each faculty is shaped entirely by experience. In the late 1700s, the French psychologist Étienne de Condillac took Locke's idea one step further; he declared that even the soul does not exist prior to experience, that it arises in response to the outside world somewhere in the brain, and presumably in accordance with a mechanical principle. Condillac's contemporary Pierre Cabanis originated a metaphor that would become somewhat notorious when he said that the brain digests sense impressions and secretes thoughts in the same way that the stomach processes food by secreting gastric juices. The senses, according to Cabanis, literally nourish the mind.

The case for nature, for the existence of innate mental dispositions, would be picked up and stated rather memorably a few years later by a young Viennese physician named Franz Josef Gall, who is usually remembered, if not blamed, for having invented phrenology. What he really invented was the first true paradigm of mind and brain physiology, one that would set the course for future neurological investigation. But it did not catch on immediately. Unfortunately, Gall also unleashed a competing paradigm that would overshadow his doctrine of the brain by shifting attention to the skull. It was only when Gall's original theory was taken up again, some thirty years after his death, that the cabinets of skulls he had made so fashionable would give way to cabinets of brains, and the search for genius could begin.

Chapter 3

Gall

WHEN MOST PEOPLE think about the shapes of heads and their contents, they think of phrenology, the discredited science of reading the bumps on the skull. During its hundred-year run, essentially the whole of the 1800s, the ridicule heaped on its inventor, Franz Josef Gall, and his successors exceeded anything now hurled at astrology, palmistry, and tarot readings. While these arts still have their followers, phrenology suffered such a precipitous fall from grace that by the time it hit bottom there was nothing left to debunk. Its premises were no longer worth mentioning, much less refuting. Or so it would seem.

But alone among the gallery of fallen sciences, phrenology did manage to get a few things right. Most people would be surprised to learn that it laid down the fundamental premise of modern neuroscience—the concept of cortical localization. Each year tens of millions of dollars are spent in the pursuit of an essentially phrenological view of brain function, which is to say in matching zones of the cerebral cortex to very specific cognitive functions. The practice still has detractors. Some critics refer to these types of investigations as the "new phrenology," which they do not mean as a compliment.

Part of the problem is the word itself. Historically, phrenology has stood for a number of competing ideas, one of which was Gall's original invention (which he did not, by the way, call phrenology—he despised the term). What Gall first proposed was a doctrine of the mind that he dubbed "organology," in which the brain is said to consist of a series of independent organs, each of which has a specialty. He then had the bad judgment to supplement this with something else called "cranioscopy," also known as his doctrine of the skull. Cranioscopy was an intriguing guess, and one that could be tested, which is what Gall set out to do. By feeling the contours of carefully selected skulls, and mapping protrusions and depressions on their surfaces, he hoped to detect the presence or absence of specific mental faculties.

After Gall, phrenology would go through several distinct phases. In its second phase, Gall's renegade assistant, Caspar Spurzheim, converted his teacher's physiological psychology into a philosophy of life. It was Spurzheim who gave phrenology its name, then took the product to America in the 1830s, where it was stripped of much of its scientific underpinnings and reborn as a self-help movement called "practical phrenology" (as opposed to Spurzheim's "theoretical phrenology"). It was in this phase that phrenology became the scam of the sideshow huckster and wandering charlatan. As Mark Twain derisively noted, the typical practical phrenologist would arrive in a new town and "furnish his clients with character charts that would compare favorably with George Washington's."[1]

In each of its three incarnations—organology, theoretical phrenology, and practical phrenology—Gall's original notion of the brain as a "congeries" of distinct organs was perceived as a threat to the moral order and was attacked relentlessly. Yet it had enough intrinsic appeal to survive all the fuss and become part of mainstream science in yet another guise—the doctrine of cortical localization. What is not clear is whether this new phrenology has legitimized Gall's original theory, or betrayed it.

FRANZ JOSEF GALLO was born in Germany in 1758 into a family of Italian descent. He studied medicine at Strasbourg, received his Ph.D. in 1785 in Vienna, where he changed his name to Gall, and began to practice medicine. By all accounts, he had a magnetic, outgoing personality, and such a knack for charming everyone (especially women) that he soon built a lucrative private practice, so lucrative that he could afford to turn down an offer to become the personal physician to Emperor Francis I. Gall preferred to retain his independence, but in a few years he would have reason to second-guess the decision.

In the mid-1790s Gall set his medical practice aside to concentrate on public lectures that drew large audiences and, with them, the attention of the authorities. His subject was not the anatomy of the brain, but the mind's relation to the brain, how mental faculties and tendencies originate in its different regions.

In 1798, Gall wrote a famous letter, in which he said, "My purpose is to ascertain the functions of the brain in general, and those of its different parts in particular; to show that it is possible to ascertain different dispositions and inclinations by the elevations and depressions upon the head; and to present, in a clear light, the most important consequences which result therefrom to medicine, morality, education and legislation—in a word, to the science of human nature."[2] This was the birth of phrenology, although the name itself would not be coined for another two decades.

Within two years the curtain would come crashing down on Gall's Vienna performance when he received a letter from the very emperor whose patronage he had declined, to the effect that: "This doctrine concerning the head, which is talked about with enthusiasm, will perhaps cause a few to lose their heads; it leads to materialism, and therefore is opposed to the first principles of morals and religion."[3] Gall had come perilously close to suggesting, as La Mettrie and Cabanis had already dared to do, that the mind is nothing more than the product of physical processes in the brain. He protested his innocence, but in vain. If thoughts can be explained by physiology and personality by anatomy, the reasoning went, then the immortal soul is a mere invention. This was blasphemy. It was time for Dr. Gall to move on.

Like most inventors, Gall did not actually invent anything. Instead he borrowed the ideas of his predecessors and combined them into something new. He did not try to hide the debt. The Swiss naturalist Charles Bonnet, who had postulated that the mechanisms of the brain might be read as the pages of a book, was one such influence. Another was Paul-Henri Thiry, better known as Baron d'Holbach, a German nobleman and strident atheist whose bravado stopped short of publishing under his own name. In his 1770 *System of Nature,* d'Holbach wrote, "Those who have distinguished the soul from the body appear only to have distinguished their brain from themselves. Indeed, the brain is the common center where all the nerves, distributed through every part of the body, meet and blend themselves . . . producing within itself a great variety of motion, which has been designated the *intellectual faculties.*"[4]

But what are the intellectual faculties? Gall thought he knew because he had seen them in action. His first epiphany occurred in grade school, when he noticed a classmate with a particularly good memory who had eyes that bulged out of his head, like those of a bullfrog. Gall made an immediate connection between appearance and personality, which was hardly a new idea. During the 1770s, the Swiss poet Johann Kaspar Lavater had tried to elevate the art of physiognomy—reading character by physical appearance—into a science. By observing facial features and body types, he proclaimed that specific shapes correlate with character traits and instinctive behaviors. In short, the outward body reveals the inner person. In Lavater's system, the face became a tripartite mirror of soul: the mouth divulging the animal tendencies, the nose and cheeks signaling moral qualities, the eyebrows and forehead indicating intellectual strength (or its absence).[5]

But unlike Lavater, Gall was not satisfied with coincidence. It seemed to him that bulging eyes were not just an outward sign of a good memory, but a symptom of it—the result of an excess of brain matter in the frontal region. With no other evidence, Gall reasoned that the part of the brain responsible for memory must lie behind the forehead. He then looked for similar correlations in the cranial shapes of people who exhibited other extremes of talent or behavior. Not surprisingly, he found them, often basing his conclusion on a single case history. (He cited Aaron Burr, who sired a child out of wedlock and had a protruding back of the skull, as evidence for a faculty of "love of offspring" in the occipital portion of the brain.)

During the 1790s, Gall studied human and animal behavior in order to reduce human thought and habits down to their essential elements. The faculties assigned by Nemesius to the ventricles—sensation, imagination, and memory—seemed inadequate to explain the variety of what he was observing. There had to be more ingredients in the mix, and Gall came up with two dozen. In his final list of twenty-six mental faculties, many of them borrowed not from Descartes but from "faculty psychology," a school of Scottish philosophy, Gall noted nineteen that are common to men and animals. These include, in addition to memory and the senses, the instincts of reproduction, self-defense, murderousness, love of offspring, friendship, cleverness, pride, vanity, and circumspection. The faculties that distinguish men from beasts, he decided, are constancy of purpose, sagacity, poetical talent, wit, religiosity, and a sense of metaphysics. He then tried to show that each of these faculties originated in a distinct part of the brain.

Gall conceded that his assignment of faculties to the seemingly undifferentiated folds of the cerebral cortex involved guesswork, that he could not account for all regions of the brain, and that his list of faculties was merely a suggestion. The map he drew was so vague that it is hardly ever reproduced in history books. His cranioscopy was, after all, merely an offshoot rather than the core of his theory. But it gained him the attention he craved.

IN HIS LAST year in Vienna, Gall took on an assistant, a medical student named Johann Caspar Spurzheim, and trained him in anatomy, philosophy, and his unique brand of showmanship. By all accounts, Gall was an expert dissector, although he did not start out as one. It was only in 1800, a full decade after formulating his theory of the mind, that he perfected Thomas Willis's bottom-up technique, and taught it to Spurzheim. In their search for anatomical evidence that would support their new theory, the two men would become the preeminent brain dissectors in Europe.

Gall's charms often obscured his talents. He was an incorrigible ladies' man, a shameless self-promoter, and a brilliant entrepreneur who scandalized his professional colleagues by charging an admission fee to his lectures. Unchastened by his eviction from Vienna, he and Spurzheim embarked on a fantastically successful lecture tour of the capitals of Europe, and their reputation grew. When they settled in Paris in 1806, no one knew quite what to make of them. According to one of his most prominent detractors, the esteemed physiologist Pierre Flourens, the good doctor was a headhunter.

> At one time everybody in Vienna was trembling for his head, and fearing that after his death it would be put in requisition to enrich Dr. Gall's cabinet. . . . Too many people were led to suppose themselves the objects of the doctor's regards, and imagined their heads to be especially longed for by him as a specimen of the utmost importance to the success of his experiments. Some very curious stories were told on this point.

> Old M. Denis, the Emperor's librarian, inserted a special clause in his will, intended to save his cranium from M. Gall's scalpel.[6]

Gall's "experiments" were fairly benign compared to Victor Frankenstein's, and they involved nothing more than collecting skulls of the dead and making plaster casts of heads of the living in order to compare them to one another. He did not preserve or collect brains. The methods for doing so did not yet exist. What really frightened the people of Vienna, and later Paris, was not what Gall might *do* to their heads, but what he might *make* of them. Gall claimed that he could feel the shape of the brain through the contours of the skull and, in doing so, assess the quality of the mind within. But that was not all he claimed.

Gall's revolutionary idea—the core of his organology—started with four principles. Each one seems reasonable enough from today's perspective, but they were profoundly and bravely original when he first proposed them. Gall insisted that: (1) the moral and intellectual dispositions are innate; (2) their manifestation depends on organization; (3) the brain is the exclusive organ of the mind; and (4) the brain is composed of as many particular and independent organs as there are fundamental powers of mind.[7] (It should be noted that Gall's doctrine of the skull—his cranioscopy—was not part of this core philosophy. Bump reading was an afterthought, and an unnecessary one, but it gained so much attention that it quickly obscured the larger theory.)

These four points, like the man himself, contained something to offend everyone. The first point irked empiricists such as Pierre Cabanis, who had originated the metaphor that the brain digests experiences in the same way that the stomach digests food. Gall's insistence on the brain as the sole organ of the mind ran up against holdouts for the humoral theory. Both Cabanis and the physiologist Xavier Bichat, among others, still adhered to the Aristotelian notion that emotions originate in the abdomen and thorax. Gall's attempt to divide the brain into a "congeries of organs," as he called it, conflicted with the unity of mind proposed by Descartes, and threatened the very idea of the unified soul. Worst of all was Gall's suggestion that the organs of the brain provide a biological basis for independent faculties of the mind. This posed a direct threat to the existence of free will, and led the Scottish philosopher Sir William Hamilton to claim that "phrenology is implicit atheism."

In Paris, resistance was immediate and came from all sides. Napoléon stepped in to prevent Gall from teaching his doctrine. On his deathbed, the emperor told his personal physician that it was the best decision he had ever made.[8] Forced to return to private practice (where his patients would include Prince Metternich, the socialist reformer Henri de Saint-Simon, and the novelist Henri Beyle, better known as Stendhal), Gall continued his dissections and produced a string of impressive discoveries. And yet, despite his reputation as a brilliant anatomist, Gall was denied admission to the French Academy. He even found it impossible to retain the loyalty of his own disciples. In 1813, after years of mounting philosophical differences, he and Spurzheim parted company after publishing the first of two volumes of their massive brain atlas, a work containing over a thousand illustrations of unprecedented accuracy. Not satisfied with making discoveries in anatomy, Spurzheim saw greater potential in Gall's doctrine of the skull than in his theory of the mind. After splitting with Gall, he elevated cranioscopy over organology. With little corroborating data, he renamed and relocated many of Gall's faculties, added a dozen to the list, produced an elaborate brain chart that fixed the boundaries of all thirty-six organs, and gave it the name phrenology, or science of the mind. Gall instantly saw it for what it was, a corruption of his theory dressed up as a personal betterment philosophy, and he wanted nothing to do with it. But in the public mind the two men were inextricably linked.

While Gall's original theory was difficult to prove, it was just as difficult to debunk. In 1808, a committee appointed by the Institut de France and chaired by the comparative anatomist Georges Cuvier reviewed Gall's work and rejected its claims to science. Although the verdict was expected, given the political climate, it was nonetheless a well-argued decision. Gall and Spurzheim had settled upon their explanation before setting off in search of evidence. Even after exploring the brain, they could provide no physiological proof that the nervous system was composed of independently functioning modules, or that any of their faculties were located where they claimed, or that they even existed. Even so, the verdict did not rule out the possibility that Gall might be right.

THE ROW OVER phrenology in France boiled down to its fundamental contention that the brain is not just one organ, but many. The idea flew in the face of several important beliefs, notably the unity of the soul, but also the facts of personal experience. No one is aware that any one part of his or her brain works harder than another, or that some kinds of thoughts come from one part of the cortex while others come from somewhere else. As phrenology's detractors would argue, consciousness is one, awareness is one, and attention is limited to one thing at a time. Clearly there are unconscious processes—the heart beats, the lungs expand, digestion occurs—all without conscious effort, and anatomists were able to locate these functions in the midbrain and brain stem. But no one can consciously handle multiple tasks without juggling them. How, then, could a brain consist of several independent, function-specific modules?

It was not until the 1820s that Georges Cuvier's protégé, a talented experimental physiologist named Pierre Flourens, designed an experiment that seemed to demolish Gall's premise. Flourens was a vivisectionist. In his most famous experiment, he opened the skulls of birds, carefully scraped away part of the cerebral cortex, sewed them back up, and observed their behavior. He found that by gradually removing swaths of the cortex, he did not eliminate functions one by one, but instead brought about a gradual diminution of all functions, up until he cut away to the minimum amount of cerebral tissue necessary to sustain life.

The dispute between Flourens and Gall is often oversimplified as a contest between strict localization and what is now called holism—the notion that the brain draws upon all of its parts to produce thoughts. Flourens did believe in localized functions, but on a more simplified scale than Gall. He proposed that the nervous system can be broken down into three essential elements: perception and volition, reception and transmission of impressions, and excitation of muscle contraction. The first of these, essentially the overriding faculty of intelligence, he assigned to the whole of the cerebral hemispheres. That was the extent of his holism. The other functions he assigned to the spinal cord and to its extension in the brain stem. Muscular coordination, he decided after studying clinical cases of brain damage, was controlled by the cerebellum, an observation that gained widespread acceptance and has been amended only to the extent that this does not seem to be *all* that the cerebellum does.

Over the course of a long career, Pierre Flourens would achieve his greatest successes as a writer of popular science, and especially as a caustic critic of phrenology and its practitioners. Spurzheim, in particular, provided him with a perfect foil. Flourens recounted one instance in which the psychiatrist François Magendie, who had served as personal physician to the celebrated mathematician Pierre-Simon Laplace, and who had preserved Laplace's brain, decided to play a trick on Spurzheim. "To test the science of phrenology," Flourens wrote, "M. Magendie showed him, instead of the brain of Laplace, that of an imbecile. Spurzheim, who had already worked up his enthusiasm, admired the brain of the imbecile as he would have the brain of Laplace."[9]

Flourens also skewered Gall (the headhunter story quoted above is one example), and yet he was also one of Gall's greatest admirers. He once wrote that he would never forget the first time he saw Gall dissect a brain: "It seemed to me as if I had never seen this organ before." To Gall he gave sole credit for having firmly established the doctrine of the brain as the organ of the mind. "It existed in science before Gall appeared," he wrote, "it may be said to reign there ever since his appearance."[10]

Both Gall and Flourens made mistakes. Gall's assumption that the skull reflects the shape of the brain, the essential premise of his cranioscopy, was known to be false even in his own day. The existence of independent faculties and their assignment to regions of the cortex never rose above the level of guesswork. Furthermore, Gall had based his locations of functions on ludicrously small samples (sometimes as small as one).

As for Flourens, he, too, was criticized for his experimental methods. Bird brains, it was later shown, are a poor stand-in for human ones, and are still not very well understood. Although he was an accomplished experimental physiologist, the ablation experiments were among Flourens's worst work: poorly conceived, badly executed, and inadequately documented. He would later enjoy many successes, including the discovery of the anesthetic properties of chloroform, but when Darwin's *On the Origin of Species* was published, Flourens, by then in his sixties, held out for creationism.

The rift between Gall and Flourens was not simply a difference of opinion. Both men knew that the sovereignty of the mind and the existence of free will were at stake. If Gall was right, Locke's blank slate, his entire case for nurture over nature would disappear. If Flourens was right, morality—the responsibility for one's actions—would still rest with the individual, and evil could not be blamed on a defective brain. The outcome would also determine the limits of scientific inquiry in the tug-of-war between the anatomists and the priests. Who would have the final word on the nature of the soul?

In the 1820s, the last decade of Gall's life, Flourens prevailed, the brain retained its mystery, and Gall's star plummeted. The struggle between the pioneering anatomist and the brilliant physiologist went to the younger man, the upholder of the status quo, and the mandarins of French academia breathed a collective sigh of relief. Forty years later, however, in the last decade of Flourens's life, a new generation of anatomists and anthropologists would resume the debate and settle it in Gall's favor. But while they resurrected his theory, they could not rescue his reputation, and they were careful to maintain a distance, if in name only, between their rediscovered science and its discredited roots.

IN HIS *Devil's Dictionary* Ambrose Bierce defines phrenology as "the science of picking a man's pocket through the scalp."[11] In "Broca's Brain," Carl Sagan calls it "a graceless nineteenth century aberration." Somewhat more charitably, Stephen Jay Gould, in *The Mismeasure of Man,* omits Gall from his rogues gallery, if only because Gall was not intent on reducing intellectual worth to a single measure. Still, it is clear that Gould's opinion of Gall's science hovers between bemusement and derision.

Anyone who is vaguely familiar with his accomplishments will appreciate the distinction between Gall the theorist and Gall the showman. Sagan admits that "there is a connection between the anatomy of the brain and what the brain does," and in that sense "there is something to the phrenological vision." Gould concedes that "cranial bumps may be nonsense, but underlying cortical localization of highly specific mental processing is a reality of ever-increasing fascination in modern neurological research." The latest edition of *Gray's Anatomy* is more direct in its praise. "Gall and his pupil Spurzheim," it says, "though achieving in phrenology a notoriety undeserved by Gall, initiated the concept of cortical localization."[12]

Which is to say that like Darwin, Gall got the big thing right and many little things wrong. But unlike Darwin, he would be held to account for the little things. Spurzheim, although a convenient foil, is not entirely to blame for this. Gall was too enamored of his cranioscopy to consider the consequences of promoting it shamelessly. His notion of bump reading was irresistible and effortlessly marketable, and although neither he nor his wayward assistant read bumps for a fee, and though they tried to restrict their readings to individuals with outstanding character traits, neither man could resist the occasional crowd-pleasing stunt, as when Spurzheim phrenologized the skull of Descartes, or when Gall visited the insane asylums and demonstrated how the shapes of the patients' heads corroborated their symptoms. By playing up its publicity value, both men exposed their project's potential for exploitation, and it soon degraded into a sideshow.

But Gall's legacy did not end there. His organological program would quietly move forward on another front in the second half of the century under another name, when the emphasis was shifted to the brain. The force behind this shift, the man who nominated Gall's doctrine of the mind as the next paradigm of the brain, was one of the unsung heroes of modern science. His name was Auguste Comte.

AT THE TIME of organology's creation, the man who would rescue it from disgrace was born into an ardently Catholic family in Montpellier, France. In 1803, at the tender age of fourteen, Auguste Comte revealed the cast of his own mind when he declared, to the shock of his parents, that he had "naturally ceased believing in God."[13] For good measure, he also denounced the monarchy, marking himself at an early age as a freethinker and republican. Not surprisingly for a youth so disaffected with authority, Comte chose to study science, then switched to engineering. At Paris's École Polytechnique he became secretary to the socialist philosopher Claude Saint-Simon (who happened to be one of Gall's patients), and through him Comte came under the spell of the ideologues, a group of social reformers that included the atheist Pierre Cabanis. But Comte would abandon science when he discovered something that Saint-Simon called social physiology. Comte would later reinvent it, and give it the name it goes by today—sociology.

As an insider in the intellectual scene in Paris in the 1820s, Comte knew about Gall and his travails, and he had some sympathy. He once wrote that "the sad state of philosophic endeavors which forced Gall to formulate his phrenological analysis in detail, later tended to discredit his conception in the eyes of serious scholars and left them to be exploited by others of limited intelligence."[14] This was a direct reference to Caspar Spurzheim and the Scottish phrenologist George Combe, whom he accused of causing Gall's vision "to degenerate into a vulgar charlatanism." In Comte's view, phrenology's compromised legacy was perched on a veneer of science too thin to support it.

Gall, of course, was complicit in his own downfall, and not above vulgar charlatanism himself. Late in his life, he visited the wards of the Salpetrière, where he made a show of diagnosing mental illnesses by feeling skulls. After the case histories were described to him, he confidently demonstrated how the form of each head confirmed the clinical diagnosis. Years later, François Leuret, who was a student at the Salpetrière at the time (studying under Spurzheim, no less), sarcastically recalled how when the tables were turned and Gall was asked to guess the clinical symptoms by feeling the head first, he suddenly fell silent. "With complete certitude," wrote Leuret, "he was able to step up from the effect to the cause, but in no instance could he descend from the cause to the effect."[15]

Leuret, who later became a pioneering psychiatrist, was not convinced that all mental illnesses arose from organic causes, as Gall contended. Comte, on the other hand, did not especially care. He wasn't bothered by Gall's antics or Leuret's objections. Both were beside the point. What piqued Comte's interest was the process by which such questions might be answered. Gall and Leuret, as he saw it, were asking the right questions, but they lacked the proper framework—not a paradigm, but a way of proceeding—that would lead them to usable answers. Comte made it his mission to provide it.

In the six volumes of his *Cours de philosophie positive,* which appeared in 1830, Comte framed his theory within the context of the historical progress of knowledge. The book contains Comte's most famous idea: that in the course of history, each branch of knowledge (by which he meant each one of the sciences) passes through three distinct stages—a theological one, a metaphysical one, and finally a fully scientific or positive one. What Comte meant is that the first humans invented mythologies to explain natural and historical events. They blamed everything on the gods. But as cause-and-effect discoveries began to supplant myths, human knowledge entered a higher stage in which it relied on metaphors more than fables. In the final, or positive, stage, metaphors gave way to paradigms, and mere rationalizations were replaced by immutable laws of nature.

How quickly a particular discipline progresses through the cycle, according to Comte, depends on the complexity of the questions it seeks to answer. Astronomy, for example, had made the passage first, and was soon followed by physics and chemistry. Biology and psychology, on the other hand, were relative newcomers.

During the second half of the nineteenth century, Comte's vision would emerge as the reigning creed of empirical science, and positivism would become its mantra. The word implies a method of inquiry based on observation, experimentation, and, where possible, comparison. Mysticism and spirituality had no place in it. Every branch of knowledge, Comte predicted, including biology and psychology, would eventually come under its grand umbrella and move into the third and final stage, where they could all be subjected to a single standard of truth. This standard would become, for Comte, the foundation of a new religion, whose saints would be the pioneering scientific visionaries. He named Gall as one of them.

Comte's positivism came along at the right time. It fed into a growing obsession with data collection and a growing dissatisfaction with metaphysics. For one scientist to call another a metaphysician, from a positivist perspective, was an insult. Anyone could make a claim, a good Comtian would say, but science requires proof in the form of hard facts, and lots of them. Metaphysicians insisted that living beings are more than the sum of their parts, that the spark of life derives from an intrinsic vital force. Positivists, on the other hand, believed that everything was potentially explainable if one looked closely enough, and in the late 1800s it was they who drove the sciences into the era of specialization.

Despite his enormous influence, Auguste Comte has become something of a forgotten man of science. As early as the 1950s, the British historian Isaiah Berlin noted that his obscurity was, in a way, the perfect measure of his success as a thinker. His ideas were assimilated so quickly and instinctively that he himself never enjoyed the credit. According to Berlin:

> Comte's views have affected the categories of our thought more deeply than is commonly supposed. Our view of the natural sciences, of the material basis of cultural evolution, of all that we call progressive, rational, enlightened, Western; our view of the relationships of institutions . . . to the emotional life of individuals and societies, and consequently our view of history itself, owes a good deal to his teaching and his influence.[16]

But Comte was not without his critics, most notably John Stuart Mill, who saw serious gaps in the positivist pyramid of scientific knowledge. Where, asked Mill, did psychology fit into the scheme? In response, Comte suggested that the third stage of psychology, the positivist stage of the science of the mind, was none other than phrenology.

By this he did not mean the "vulgar charlatanism" of Combe or Spurzheim, or skull collecting and bump reading. Instead he proposed a return to Gall's original four points, to his organology, a scientific paradigm that, as he wrote to Mill, "was not carried out in sufficient depth and was not accomplished as it should."[17]

From today's perspective, Comte looks like a visionary. "[T]he brain," he wrote, "is no longer an organ but an apparatus of organs, more complex in proportion to the degree of animality."[18] To study it scientifically, Comte suggested, a new breed of phrenologists "must make a much more extensive use than hitherto of the means furnished by biological philosophy for the advancement of all studies related to living bodies; that is, of pathological and yet more comparative analysis." In other words, the path of progress in understanding the brain lay in morbid anatomy—the correlation of symptoms with brain lesions—and comparative anatomy—the comparison of brains of all species.

All that was required was the material itself. Gall's skull collections and his phrenological busts were no longer of any use. Only brains would do. But there was a problem. Acquiring them was easy enough. The real trick was to keep them from rotting away.

Chapter 4

Byron

IN 1824, EIGHT years after having suggested to Mary Shelley that she write the ghost story that became *Frankenstein*, Lord Byron died in Greece, where he was hailed a hero of the war of independence. At that time, the postmortem examination of corpses (as opposed to mere embalming) was something new. German physicians had been promoting the practice as an educational tool with some success, although there were as yet no standards in place, particularly for the disposition of organs. Whenever a truly important person died, it was not uncommon to preserve his or her heart, and to accord it a separate burial, because the heart was still regarded, if only sentimentally, as the seat of the soul.

Byron was well aware of the practice. The day before he died he said to his doctor, "One request let me make to you. Let not my body be hacked or be sent to England."[1] But the request would not be honored. When the attending physician prepared the corpse, he removed Byron's brain, recorded a weight of "about six medicinal pounds," and placed it in an urn. In two other urns he placed the heart and the viscera. After much discussion, the coffin and urns were shipped to England to be viewed (by Mary Shelley, among others), then interred in the Byron family plot at Hucknall, outside of London. The brain rotted away, but news of its extraordinary size soon began to spread.

PRIOR TO THE 1800s, the preparation of corpses was left mostly to specialists—embalmers and undertakers—which presented anatomists and physicians with a problem: how to study an object that was, by sacred law, off limits to them. Cadavers had always been hard to come by. The sole exception to this rule seems to have occurred in ancient Alexandria, where the first anatomists of record, Herophilus and Erasistratus, dissected over a thousand bodies. According to the ancient chronicler Celsus, "They laid men open while alive—criminals received out of prison from the kings—and whilst they were still breathing observed parts which beforehand nature had concealed."[2] In the process, Erasistratus described and named the cerebrum, the cerebellum, and the ventricles.

This gruesome window of opportunity then closed, but not completely. Conceding the necessity of autopsies as a teaching tool, as a means of gathering forensic evidence, and often merely as entertainment, rulers doled out corpses very carefully, selecting them from the dregs of society. In the thirteenth century Frederick II, the holy Roman emperor, provided two executed criminals every other year to the medical schools. The Council of Edinburgh, in 1595, did much the same. The famous anatomy theater in Padua hosted infrequent public dissections that generated the kind of frenzy that now attends heavyweight title fights. Criminals, again, provided the fodder, but the hardest ticket to get was on those rare occasions when a woman's body was to be cut open. In seventeenth-century France, dissections were legalized by royal decree, but the executioner could not keep up with demand. A French physician, making a virtue of necessity, noted, "My friend was ill. I attended him. He died. I dissected him."[3]

This state of affairs would hold the progress of anatomy in check up to the dawn of the twentieth century, with occasional exceptions. During the mid-1600s, Thomas Willis was allowed to perform autopsies (including brain removals) on members of the gentry. This was partly due to the esteem enjoyed by natural philosophy at the time, but mostly because the Anglican Church took a far less proprietary interest in the human body than did the Church of Rome. Willis and his colleagues took full advantage. It is no coincidence that this opportunity produced the greatest breakthroughs in brain anatomy since the time of Galen. But England soon returned to its old ways, and by the 1800s, the standard practice among "professors" of anatomy—which is to say, among unaffiliated instructors who ran anatomy schools out of their houses—was to buy corpses or to steal them. In doing so, they stimulated a brisk trade in the recently dead.

The term *grave robbing*, it should be noted, refers to the practice of digging up coffins in order to steal the jewels and other valuables in them. *Body snatching*, on the other hand, became increasing common when bodies themselves became valuable as commodities, and thieves realized that there was no law against taking them.[4]

The lore of body snatching is a rich one. It was a source of dread among the newly bereaved, especially during the eighteenth and nineteenth centuries, and the lengths to which some people went to protect their loved ones' remains were exceeded only by the robbers' stratagems for foiling them. The rich built their mausoleums like fortresses. The poor had to resort to vigilance. Anxious friends and relatives would often guard gravesites for days, sometimes in vain. Many bodies never made it to the cemetery. For a small fee, undertakers were happy to use sand to bring an empty coffin up to weight. For a time in eighteenth-century England, children's bodies were sold by length. The practice was illegal, and more than a few prominent physicians went to jail for it. The notorious Edinburgh body snatchers William Burke and William Hare even resorted to murder to keep up with demand. They were sentenced to die by hanging, although their biggest client, a renowned anatomy professor named Dr. Robert Knox, was merely accused of willful negligence. Twenty-five thousand people turned out to watch Burke hang, and an equal number filed past his dissected cadaver at a local laboratory. His skeleton can still be found at the Royal College of Surgeons in Edinburgh, along with several purses made from his skin.

Given Burke's fate (the inspiration for the epithet "Burke him!"), it is not surprising that dissection was considered a fate worse than death. It was one thing to be sentenced to die at the gallows, but to be sentenced to die and then be publicly flayed was appalling, especially given the official Church doctrine that souls retain the appearance of their dead selves in the afterlife. What would make postmortem examinations acceptable was a collusion of mortuary and anatomical sciences. The fear of dissection was mitigated when anatomists learned how to open bodies in such a way that they could return them to a natural state. By 1824, it was not shocking that Byron would have undergone an autopsy, or that his heart would have been removed. It was by then a fairly standard procedure, and he expected it. But for his brain to be taken from his skull and weighed by a local physician in a small Greek town was unusual, if not unprecedented.

The size of Byron's head probably had something to do with it. Depending upon the system of weights employed and the interpretation of a "medicinal pound," the brain might have weighed anywhere from 1,807 to 2,238 grams (the lower and more likely figure being a conversion from Venetian pounds; the higher, from English pounds, would make it the heaviest brain on record). In any case, it was a massive brain, at least 25 percent bigger than the average, and the connection with his outsized reputation was soon made. Byron, the poster boy for romantic genius, became Exhibit A for the contention (still widely held, although long ago disproved) that great intellects are housed in large brains. It was just the beginning.

THREE YEARS LATER, in 1827, Ludwig van Beethoven died in Vienna, and his physicians removed the temporal bones of his skull, hoping to find an explanation for the composer's deafness. Instead they turned up yet another curious fact. Although they merely exposed the brain and did not remove it, the pathologist, Dr. Johann Wagner, could not help noticing that "the convolutions appear twice as numerous and the fissures twice as deep as in ordinary brains."[5] A second criterion now fell into place. It seemed that superior brains not only weighed more, but they were more richly convoluted and deeply fissured than average brains. Not that this made much of an impression at the time. What did leave an impression with viewers of Beethoven's coffin, although not a pleasant one, was his bluish, misshapen countenance. It is not easy to restore a face when a good deal of the skull is missing, and a wreath of white roses encircling the composer's brow did little to hide the crude handiwork of the pathologists.

As the heart gave way to the brain as the seat of the intellect, the prospect of having one's gray matter removed became, at least to some minds, an honor rather than a sacrilege. The great mathematician-astronomer Pierre-Simon Laplace, who once served as Napoléon's tutor, probably knew what his physician was thinking as his life slipped away. François Magendie, a pioneering anatomist, removed and preserved Laplace's brain in 1828, then kept it on his desk. It is one of the first recorded instances of a preserved brain.

Franz Josef Gall also died in 1828; he was denied the sacrament of last rites, as he would have expected, having been excommunicated in 1817 by Pius VII, when his works were placed on the Index of Forbidden Books. He once protested that he had never questioned the nature of the soul. "In studying the works of God, I do not think I have done any wrong," he insisted, but in vain.[6] His reputation suffered even further when his brain was removed and weighed at autopsy. Not only did his skull turn out to be twice as thick as an average skull (providing more fodder for his critics), but his brain weight, recorded as 1,312 grams, was reduced through subsequent transcription errors to a mere 1,192 grams (a figure reproduced in *The Mismeasure of Man* by Gould, who called it "a crowning indignity").

As for Gall's former partner, Caspar Spurzheim, he would die four years later, and his brain would serve the same decorative purpose as Laplace's. Upon his arrival in Boston in 1832, he was received like a prophet, and he soon became the toast of the town. But his social obligations took their toll. After delivering a series of sold-out lectures at Harvard, he suddenly took ill and died of pneumonia exacerbated by overwork. Thousands turned out for the funeral procession, which proceeded to the prestigious Mount Auburn Cemetery, where the body was laid to rest, sans head. It had been removed, and both the brain and skull were kept by the American Phrenological Society. For years the brain graced the desk of the director of the Boston Athenaeum. It has since disappeared. The skull can still be found, along with the rest of Spurzheim's phrenological collection of skulls and plaster busts, at the Countway Museum of Medicine at Harvard.

Before the year 1832 was out, the great French naturalist Georges Cuvier would die in Paris at the age of sixty-three. Cuvier had pioneered the field of comparative anatomy, explored the fossil record, and helped to establish Paris's preeminence in natural science during the nineteenth century. He had also chaired the commission that in 1808 investigated Gall's organology. Cuvier might have been gratified to know that his own brain was huge, a practically Byronic 1,830 grams. Yet oddly, it, too, was not preserved, nor was Cuvier's skull, despite the fact that ten prominent physicians attended the autopsy. But as in Byron's case, the size of the great man's head would not go unnoticed.

IN HISTORICAL TERMS, Byron's brain was an oddity, as was Cuvier's. The size and shape of their brains turned out to be a happy accident, one that was soon freighted with great significance. Byron's brain, for example, would inspire the notion that genius is a matter of brain power, and that prolific dead white men possessed the most well developed brains of all. Cuvier became a symbol of Gallic superiority. On the other side of the ledger, the small skull of the Venus Hottentot, an African woman who performed in a traveling circus and whose skeleton was preserved at Cuvier's museum of natural history, would stand in for an entire continent, and epitomize for a while the erroneous notion that the brains of whites are better developed than those of blacks.

Clearly there was a need for more data, and over the next four decades anatomists would begin to supply it. Some of them even volunteered their own brains. As early as 1815, the German anatomist Friedrich Tiedemann was the first to offer his brain for scientific study (a promise he would not have to honor until 1864).[7] In the years after Byron's death, improvements in preservation techniques made such bequests possible. The earliest case on record is that of Ignaz Döllinger, a prominent professor of anatomy in Munich who died in 1841. Döllinger left detailed instructions with his students for the disposal of his body, and his brain was preserved. If it was not the first brain to receive that dubious honor (Laplace and Spurzheim had preceded him by a decade), Döllinger's was the first brain to be preserved explicitly "in the interest of science." The question was, what would science make of it?

IN ORDER TO make anything of a brain, in order to handle it without damaging it, it has to be stabilized. Ideally, its cells should be fixed in place for all time, which is more easily said than done. Thomas Willis and Nicolaus Steno relied on such half measures as boiling in brine, soaking in oil, and marinating in wine—all of which postponed the onset of decomposition and bought them enough time to make their slices. But for archival purposes, what they needed was a way to keep wet specimens wet and intact for an indefinite period.

In 1789, Marie Antoinette's personal physician, Félix Vicq d'Azyr, devised a solution of alcohol mixed with salts that preserved tissue indefinitely.[8] (He did not preserve the queen's brain, by the way, being too occupied with escaping the guillotine himself.) But the first successful preservation of a whole brain in alcohol did not take place until two decades later, and did not become widely known for a few decades more. The trick was discovered in 1809 by the German anatomist Johann Christian Reil, and it would continue to be used until the remarkable penetrative powers of formaldehyde were discovered by Ferdinand Blum in 1893. Formaldehyde immediately became the fundamental ingredient in all tissue preparation solutions, and it has remained so ever since.[9]

While it might seem simple, there's more to preserving a piece of tissue than plunging it into an alcohol or formaldehyde solution. The goal, after all, is to be able to study macroscopic and microscopic structures. The problem with alcohol is that it dries the tissue, causing it to shrink and sometimes to develop cracks. Moreover, its penetrating powers are limited. When a specimen as large and as dense as an adult brain is submerged in it, the surface will harden while the core deteriorates into an amorphous pulp.[10] The trick is to stave off decomposition while maintaining the brain's shape and size, as well as the integrity of its cells, and the best way to do this is first to inject a saline solution into the arteries to flush out the blood, and to follow up with an injection of preservative fluid just before removing the brain, or immediately afterward. No technique is perfect, there is always some shrinkage, and although formaldehyde remains the basic ingredient of all fixatives, no single preservation technique (all of which involve successive chemical baths over the course of several weeks, if not months) has emerged as a standard. The result was (and remains) a lack of a uniform method for pickling brains, making it impossible to compare brains prepared by different examiners.

Once a brain has been removed and stabilized, it can be examined. But problems arise here as well. With the development of powerful microscopes in the 1840s, methods for staining and slicing tissue also had to be invented, and with them came techniques for embedding blocks of brain tissue in paraffin and celloidin. In order to mount the tissue on microscope slides, the brain was first hardened and cut into numbered blocks that were embedded in wax (paraffin) or plastic (celloidin). The chunks were then sliced on a microtome, a device with a long fixed blade (essentially a meat slicer), then mounted on glass and stained with one of several dyes. They were then ready for inspection.

The first investigator to use a compound microscope powerful enough to reveal neural structures was Jan Evangelista Purkinje, a Czech physiologist who in the 1830s uncovered the basic facts of the cell theory, including the cerebellar neurons that were subsequently named for him (Purkinje cells). Although the cerebellar neurons stood out clearly enough, Purkinje could not see the neurons of the cerebral cortex because, in the absence of sophisticated staining processes, they are effectively invisible. Even the earliest stains, discovered in the 1840s and 1850s, could not reveal them. In 1840, the Frenchman Jules Baillarger mounted thin slices of cortex between glass plates, and by holding them up to a strong light, he was able to make out six cell layers. But it was not until 1872, when an Italian anatomist named Camillo Golgi discovered an ingenious staining method, that it became possible to make a selection of individual neurons come into focus. Golgi's stain acted like a photographic emulsion to bring forth an amazing wealth of detail.

Such discoveries were facilitated by new methods of tissue slicing that achieved the state of the art in 1875, when Bernhard von Gudden invented the microtome for sectioning entire hemispheres. By century's end, there were dozens of methods of preserving, staining, slicing, and mounting nervous tissue, and most researchers either adopted the methods of their teacher, or chose what appeared to be the most sensible method. What was missing was a standard protocol for preparing brains for study, and this situation remains true, to a lesser extent, even today.

THE MOST OBVIOUS thing to do with an extracted brain is to weigh it. But as Harvard neuroscientists Percival Bailey and Gerhard von Bonin caution, "to weigh a brain is not as easy as it sounds."[11]

In their influential work *The Isocortex of Man* (1957), Bailey and von Bonin point out that lists of brain weights are deceptive. Given the number of uncontrolled variables involved, such "facts" as the weight of Byron's brain, or Gall's or Cuvier's, are useless. In his 1965 survey of brain-weight studies, the South African researcher Phillip Tobias listed all of the factors that make brain weights highly suspect, if not unreliable as statistical evidence. First there is the method of removal. Where, exactly, did the pathologist sever the spinal cord? Did he allow the cerebrospinal fluid to drain out of the ventricles or not? In cases of cerebral hemorrhage, was the blood drained before weighing? There are other factors that have to be taken into account: the age of the deceased, the cause of death, general health and nutrition, childhood diseases, lifestyle, climate—all of these can potentially affect the growth of the body, and since body height is loosely correlated with brain weight, anything that affects growth, especially diet, will play a role in the brain's development.

The source of the brain can also skew the data. Historically, career criminals are overrepresented in brain studies. As a group, they tend to die young and very suddenly. As a result, their mean brain weights tend to be higher than those of other groups, especially those of elderly scholars. Women's brain weights average about 10 to 15 percent less than men's, a fact accounted for by the same average disparity in body size. Consequently, women's brain weights should not be averaged in with men's. After criminals, perhaps the most common source of brains is the poor and indigent, whose unclaimed bodies often wind up as cadavers. In a 1927 essay entitled "A Liter and a Half of Brains," Western Reserve University professor T. Wingate Todd noted that since 1913, when the city started keeping records, the average brain volume for adult males drawn from "Cleveland's outmaneuvered victims in the struggle for existence" fluctuated in lockstep with certain economic indicators, notably unemployment. "The pneumonia of the shiftless, the tuberculosis of the over-wearied struggler, the heart disease of the adventurer no longer acted alone as our receiving agents," Todd rather grandiloquently noted (he was giving an after-dinner speech to a group of doctors). "Instead men shot themselves, or each other, threw themselves in the lake; perished of cold, listlessly lost in despair."[12] As they did so, the average brain size slowly crept upward.

In the mid-1800s, most researchers relied on the volume of the skull as a dependable indicator of relative brain size, but there were problems here as well. Cranial capacity, it turns out, does not correlate perfectly with brain weight. Some brains fill more of the skull than others. As Todd noted, "of two heads of the same size one might have as much as 200 cc of brain more than the other." This fact, unknown to the phrenologists and craniologists of the early 1800s, would make most of their data collecting useless.

What everyone wants to know, of course (if they do not already assume it), is whether the size of the brain is related to intelligence. This, as will be seen, is a more complicated question.

AFTER WEIGHT, the most obvious point of comparison between brain specimens is appearance, specifically the complexity of fissural patterns, the development of the convolutions, and the relative sizes and degree of asymmetry in the lobes. Here, too, the problems of comparing brains are daunting.

In his early studies of nervous tissue, Félix Vicq d'Azyr was particularly frustrated by his inability to describe what he saw in scientific terms. "Seeing and describing are two things that everyone believes he is capable of doing," he wrote, "yet few people can really do them."[13] The problem was a lack of a specific terminology "necessary for conveying an exact idea of what one has seen." The parts of the brain, in particular the landscape of the cerebral cortex, was still an uncharted territory.

This problem would be solved largely through the efforts of two men: François Leuret, the chief psychiatrist at Paris's largest insane asylum, the Bicêtre, and his talented assistant, the comparative anatomist Louis-Pierre Gratiolet. In their two-volume work, *Comparative Anatomy of the Nervous System Considered in Relation to Intelligence,* which they labored over between 1839 and 1857, the two men identified the common features of all primate brains and named the principal fissures, convolutions, and folds. (Leuret died in 1851, leaving it to Gratiolet to complete the second volume.) Much of this work was facilitated by the perfection of the alcohol-preservation techniques described above, which allowed Gratiolet to study the cortical surface of his specimens at his leisure, to find structures that had never been properly observed before, and name them.

By 1850, with methods for removing and preserving brains firmly in place, and with increasing opportunity to study interesting brains, a new line of investigation was thus poised to begin. In addition to the brains of Byron, Beethoven, Gall, Spurzheim, and Döllinger, the list of early case studies included the brains of Guillaume Dupuytren, the surgeon who had removed Cuvier's brain, the Scottish physician John Abercrombie, the Italian composer Gaetano Donizetti, and the American statesman Daniel Webster.

Other than the size of Byron's and Cuvier's brains, nothing suggested that there was such a thing as an anatomy of genius. This was very much a new line of investigation, and individual brains offered intriguing clues, but nothing had proved the phrenological contention that size and organization mattered in a very specific way. What was missing was a cabinet of exemplary brains that could be compared down to the smallest detail, and putting such a collection together would require a great deal of luck.

It so happened that, in 1855, an enterprising physiologist named Rudolf Wagner got lucky. He managed to acquire a brain so legendary that it opened the door to more acquisitions. Within a few years Wagner would remove and preserve the brains of several highly educated men under identical conditions. If an anatomical basis for their intellectual achievements existed, surely Wagner would find it.

Chapter 5

Gauss

AT THE DAWN of the nineteenth century, Germany could lay claim to three of the greatest geniuses who ever lived: one in letters, one in music, the other in the sciences. From 1800 to 1820, Ludwig van Beethoven, Johann Wolfgang von Goethe, and Carl Friedrich Gauss were at the height of their powers. Of the three, Gauss is easily the least well known, but his legacy still permeates the flow of everyday life, from the unit of electromagnetic intensity named for him, to the bell curve (also called the Gaussian curve), to the statistical method of least squares, which allowed him to plot the orbits of the planets. The list goes on, although for sheer cultural impact it contains nothing that can rival *Faust* or the *Eroica Symphony*. To place Gauss in proper perspective, it would be more fitting to group him (as most historians do) with Sir Isaac Newton and Archimedes as the greatest trio of mathematicians who ever lived.

To appreciate Gauss's impact, a comparison might prove useful. While he is not considered a paradigm maker of the same order as Einstein, the parallels between the two men's lives are intriguing. They were born almost a century apart—Gauss in 1777, Einstein in 1875. With their breakthrough doctoral theses, each man established himself as a world-class scientist while in his early twenties, then proceeded to pile one important discovery on another. At midcareer, both men settled down in prestigious university towns that have since come to be associated with them—Gauss in Göttingen, Germany, and Einstein in Princeton, New Jersey. In a crowning irony, when they died, also a century apart—one in February of 1855, the other in April of 1955—their brains were removed and preserved in the interests of science. Or so it was claimed. In Einstein's case the claim was dubious, but Gauss was different. Byron's may have been the first famous brain to have been removed and weighed, but Gauss's was the first to be examined by a highly credentialed expert, and matched up against the brains of a few of his eminent peers. A hundred and fifty years after Gauss's death, these remarkable brains can still be viewed in the town where these men lived, worked, and died. Although not featured on any tourist maps, they are well worth a look, if only because they form a direct link to Einstein's brain. Which is to say, these were the brains that kicked off the search for the anatomy of genius.

CARL FRIEDRICH GAUSS was born into a working class family in Braunschweig in northern Germany. His father, a laborer and sometime masonry foreman, noticed his son's facility with numbers before the boy could even talk. In his first class in arithmetic, Gauss stunned his teacher by finding the sum of a hundred consecutive numbers, seemingly off the top of his head. There is a trick to it, of course (just add the first and last numbers and multiply by fifty), and finding the trick would prove to be Gauss's special gift throughout his life: he derived the proofs of many fundamental theorems before encountering them in books. At times, it seemed as though the young Gauss was rediscovering the history of mathematics from scratch. When a useful method was not to be found in the literature, he supplied it. When a theorem was glaringly necessary, such as the uniqueness of prime factorizations, he proved it. According to some accounts, by his late teens Gauss was making discoveries so fast that he could not write them all down.

Although Gauss also excelled at languages, and even considered becoming a philologist at one point, he cast his lot with mathematics at the age of nineteen when he solved a problem that had eluded geometers for over two thousand years—a compass-and-straightedge construction of a seventeen-sided regular polygon. His breakout year was 1801, when he published a collection of papers under the title *Disquisitiones arithmeticae*, which quickly joined the canon of mathematical literature alongside Newton's *Principia*. Gauss had taken on nothing less than a survey of the history of number theory and algebra, with the goal of tying up its loose ends. It was an astounding piece of work for someone so young. But while the book brought him the acclaim of the mathematical elite (admittedly a small and highly select group), it was Gauss's astronomical discoveries that would bring him worldwide fame.

In January of 1801, an Italian astronomer named Joseph Piazzi discovered what he thought might be a comet, or possibly a new planet, hovering just above the horizon. After he logged a series of coordinates, the object disappeared from the winter sky. Once Piazzi published his data in an astronomical journal, the game was on: where and when would Ceres reappear?[1]

Sometime during the summer of 1801, Carl Gauss happened on the article, which described Ceres as "a long supposed, now probably discovered, new major planet in our solar system between Mars and Jupiter." He was instantly intrigued by the geometrical beauty of the problem. Orbits, it turns out, are not perfectly circular. To confuse matters even more, astronomical sightings are not always reliable. There is always a degree of error in the recorded data, and Gauss understood the nature of measurement errors better than anyone. A few years earlier he had worked out a method (later to become known as the method of least squares) for calculating the best guess among a set of conflicting measurements. As an astronomer, Gauss's greatest advantage was that he did not think like an astronomer. Using techniques from algebra and geometry, he spent several months searching for a general solution to the problem of sightings and orbits, then applied it to Piazzi's data, and waited for Ceres to reappear.

The story has a predictable ending. Among those vying to solve the orbit, only Gauss had it right, and almost overnight, the brilliant but obscure mathematician became an international celebrity.

In 1807, Gauss was appointed director of the observatory at the University of Göttingen, and in the decades that followed, he ventured into physics, statistics, geodesy, and other math-related fields. He was no dabbler. He made original and groundbreaking contributions to every field he entered, and he continued his prolific output well into old age. With his friend, the physicist Wilhelm Weber, he even designed and built the first electromagnetic telegraph in 1833, some four years before Samuel Morse. Gauss devised a binary code, strung a wire across the Göttingen campus, and exchanged daily greetings with Weber. When he pitched the idea to the German railway industry in 1834, they failed to see its potential, leaving the field open to Morse, who applied for and won his patent in 1837.

By the 1850s, Gauss was a legend. When the university feted him at his golden jubilee, the fiftieth anniversary of his doctorate, he was the toast of Europe, and the only survivor of the great triumvirate of geniuses. But no one knew quite what to make of him. He was the prototype of the lone genius—a solitary, almost mystical savant, who in old age became increasingly isolated. His wife had died. His favorite son, a mathematical prodigy, had emigrated to America. He carried on his closest friendships by correspondence. He lived the last forty-seven years of his life in modest quarters at the observatory that had been built for him when he accepted the job at Göttingen in 1808, and, during this time, it is said, he rarely spent a night away from his own bed. He was an oddly private man, he rarely collaborated, he had few students, and he left behind no school of thought. It would take another fifty years for his fellow mathematicians to appreciate fully the breadth and depth of his output. Much of it was buried in unpublished notebooks. In the last few years of his life, his friends and colleagues knew only one thing for certain—Gauss was a genius. If ever a brain *had* to be different, it would have been his.

THE VERY IDEA of a "genius brain" was unthinkable prior to the advent of scientific materialism, a radical theory which had been steadily gaining ground since the 1820s. By the mid-1850s it was well on its way to becoming the reigning creed of the natural sciences, particularly in physiology, where it had created a generational rift between a very embittered old school and an outrageously confrontational new one.

Materialism was the natural offspring of Réne Descartes's mechanistic explanation of the human body, without the metaphysical pretense of a ghost in the machine. It was advanced by La Mettrie, Cabanis, d'Holbach, and even Gall, to the point that by 1850, questioning the existence of the soul had ceased to be a capital offense. Yet it was still an extremely polarizing issue, as much a political hot button as a religious or even scientific one. In Germany, when a group of young materialists began to challenge their older colleagues to defend their antimaterialist views, the confrontation turned into a public spectacle. At the center of the maelstrom were two men of very different dispositions—Rudolf Wagner, a staunch representative of the status quo, and Carl Vogt, a young radical and freethinker hell-bent on bringing down the establishment. To do it, he would first try to bring down Rudolf Wagner.

Wagner had impeccable credentials. He had studied under the great comparative anatomist Georges Cuvier in Paris, then worked his way up the academic ladder, reaching the top when he arrived at the prestigious University of Göttingen in 1840. During the early years of his career, Wagner's research interests included the microscopic structure of mammalian ova and sperm. He also made important discoveries in the circulatory and nervous systems, and shared in the discovery of the tactile corpuscles of the skin. In one groundbreaking study, he explained the electric organ of the torpedo fish. During the 1840s, Wagner also edited the comprehensive *Handbook of Physiology*, the bible in his field. But his string of successes ended dramatically in 1845, when at forty years of age Wagner suffered a serious heart attack that, as often happens, caused him to reassess his life's work. From that point on, he began to indulge in philosophical speculations, with special emphasis on the existence of the soul. It was this midlife crisis that would drag him into the greatest conflict of his career.

Rudolf Wagner could have spared himself a great deal of trouble if he had simply steered clear of Carl Vogt. Twelve years his junior, Vogt railed against everything that Wagner valued. Born in 1817 in southern Germany into a family dominated by a long line of clerics, Vogt had rebelled at an early age. Tossed out of one school after another, he settled down only after discovering science. First it was chemistry, which he studied with Justus von Leibig, one of Europe's preeminent scientists, then anatomy and zoology. He spent a few years with Louis Agassiz, exploring the fossil record in the glaciers of Switzerland, and laying the geological groundwork for challenging the Bible's account of the creation and the flood. But the headstrong Vogt, ever suspicious of authority, had one run-in too many with the equally headstrong Agassiz, who had the bad habit (still prevalent in academia) of placing his name atop publications written mostly by his students. Vogt broke with Agassiz in 1844 and left for Paris, where he met the ailing Russian radical Mikhail Bakunin, supposedly dying of cholera. Vogt arrived in time to deliver the news that it was merely a bad case of indigestion, and a friendship was struck.

During his Paris years, Vogt paid his rent and honed his rhetoric by writing science articles for the popular press. With friends like Bakunin, his views became increasingly politicized, and they would remain so when he drifted back to Germany. He settled into a spate of respectability as a professor of zoology at Giessen, but it would be short-lived. His support of the radicals in the failed revolution of 1848 would cost him his professor's chair and force him to flee the country.

When Vogt is remembered today, it is not for his association with Louis Agassiz or for his numerous discoveries in paleontology. Instead he is identified with a small coterie of radicals who used scientific arguments to undermine the foundations of both church and state in the mid-1800s. Vogt insisted, in rhetoric bordering on the fanatical, that physical laws can explain all things in the universe. It is difficult today to appreciate just how threatening the idea was, but in the hands of young radicals like Vogt, it became the engine that would propel modern science out of the shadow of religion and superstition, while driving the church out of public institutions (at least in France). Among his immediate group (the others were Ludwig Büchner and Jacob Moleschott), Vogt was easily the most accomplished scientist, the least consistent thinker, and the greatest popularizer of radical ideas. He was also theatrical, confrontational, and outrageous enough to lure Rudolf Wagner out from the protective walls of his ivory tower, into an unwinnable fight.

LIKE MOST PHYSIOLOGISTS of the era, Wagner was a vitalist. He was confident that science could explain every phenomenon within its sphere, but he saw it as his moral duty to limit that sphere to observable facts. According to the vitalists, life cannot be reduced to finite mechanical processes. It is more than the sum of its tangible parts. The undetectable, but nonetheless real piece of the puzzle is a vital force that infuses all living things, acting like an invisible fluid. Its power to animate flesh is the pretext of Victor Frankenstein's experiment in Mary Shelley's novel, and its inexplicability is what separates men from beasts, if not from machines.

Wagner was not alone in this belief. Among his fellow scientists, this was the majority view. His predecessor, the renowned anthropologist Johannes Friedrich Blumenbach, had subscribed to the same paradigm, as did Johannes Müller, Germany's preeminent research physiologist of the 1820s. In fact, through the 1840s most anatomists and physiologists clung to the idea in one form or another. But it was Müller's own student, Hermann von Helmholtz, who would take a sizable chunk out of the vitalist edifice in 1850, when he succeeded in measuring the speed of nerve impulses (a surprisingly slow twenty-seven meters per second, as it turns out). After Helmholtz (whose own brain would eventually be harvested for study), what had previously defied description suddenly seemed explicable. The nervous system, far from acting on vital spirits or mysterious essences, did indeed seem to act like a machine.

If the vitalists were not quite a united front (and they were not), their materialist opponents were even more difficult to pin down. While trying to eliminate religion from public life, they did their best to imitate the disciples of a new church: they quickly divided into sects. Deistic materialists conceded the existence of a creator, but one who merely set the universe in motion and then sat back to watch. (He was allowed to intervene now and then—thus the miracles.) Emergent materialists argued that at the level of brain cells and thoughts, a different, as-yet-unknown, set of physical laws comes into play. Instead of a ghost in the machine, they proposed a high-level and somewhat mysterious mechanical process. At the far extreme, strict mechanical materialists, sometimes known as reductionists, staked out the fundamentalist position. As with all extreme positions, this one attracted its share of troublemakers at first. Today it has become the mainstream position among brain scientists (although the concept of "mechanics" has evolved considerably).

Carl Vogt was a reductionist. He conceived of matter in purely Newtonian terms as consisting of indivisible particles that followed the same rules of action and reaction as billiard balls, resulting in a universe that unfolds like a stupendously elaborate trick shot according to an unalterable plan. In Vogt's scheme, a man had no more free will than an eight ball.

In the mid-1800s, philosophical differences between vitalists and materialists were not so contentious that the two sides could not work together to solve practical questions, but Wagner and Vogt were exceptions to this rule. Within the vitalist camp, Wagner staked out one extreme by asserting the existence of an immaterial soul substance that infuses the living body and flees at the stroke of death. Vogt, on the opposite extreme, ruled out any basis for the soul, and in so doing flaunted his contempt for Wagner.

"Every natural scientist," he wrote, "who thinks with any degree of consistency at all will, I think, come to the view that all those capacities that we understand by the phrase 'psychic activities' are but the functions of the brain substance."[2] As a savvy provocateur, Vogt perfected a writing style that headed off all attacks before they could be launched. "To assume that the soul makes use of the brain as an instrument with which it can work as it pleases," he wrote, anticipating the objections of Cartesian dualists, "is pure nonsense." Vogt had no compunction about stealing from his fellow atheists. His most notorious quip—"To express myself a bit crudely here, thoughts stand in the same relation to the brain as gall does to the liver or urine to the kidneys"—was a crib from Pierre Cabanis that he reworked as: "We are in no ways masters of ourselves, any more than we are masters of what we secrete and don't secrete." This infamous remark, eventually distilled as "the brain secretes thoughts in the same way that the liver secretes bile," would inspire Karl Marx and Friedrich Engels to refer to Vogt's mechanistic views as "vulgar materialism," and to respond with a materialism of their own, a more sweeping interpretation of human history that would come to be known as dialectical materialism.

In 1854, Wagner wrote an editorial for a Frankfurt newspaper in which he attacked Vogt's uncompromising views. At the time, Vogt was a professor of geology at the University of Geneva, and a respected physiologist in his own right. Most of Wagner's colleagues opposed Vogt's brand of materialism, but not always for religious reasons, or even scientific ones. What irked them even more than the theory itself was the very idea of public debate. Among German academics, the press was not considered a suitable forum for intellectual exchanges, so Wagner was taking a calculated risk when he decided to confront Vogt in the pages of a popular journal.

The feud was more tactical than ideological. Once the battle was joined, Wagner presented his case before a scientific conference he had organized in September of 1854, purposely choosing a date when neither Vogt nor his supporters could attend. In his talk, entitled "The Creation of Man and the Substance of the Soul," Wagner defended his two dearest beliefs: the descent of all humans from a single pair, and the existence of an ethereal soul substance. The first claim, known as monogenism, follows the Bible's account of the creation. The second was a form of vitalism that dated back to Descartes. Having no evidence to support either claim, the best Wagner could hope to do was to line up some influential allies. But only one of his colleagues would have anything to do with it. This was the elderly Carl Gauss.

IT WAS LARGELY out of desperation that Wagner visited Gauss. They met four times at the Göttingen Observatory during the last two months of 1854, and after each of their meetings, Wagner would rush home to write down everything Gauss said as well as he could remember it. But whether his memory was faulty, or his narrating skills weak, or simply because Gauss was evasive, the result is a scattershot portrait of the man and a nearly incoherent discussion of the issues burning in Wagner's mind.

After Gauss's death, Wagner tried to publish an account of the conversations, but Gauss's family and friends suppressed it. It only came to light in the 1950s when it was published as "Conversations with Gauss During the Last Months of His Life." The best that can be said is that it's a strange document. At each turn, intriguing questions go unanswered. The two men touched on many issues, including the mathematical foundations of psychic phenomena (too difficult to measure, said Gauss), mathematical ability (Gauss thought it owed as much to inner drive as to talent), the philosophy of mathematics, Gauss's admiration for Newton and Johannes Kepler, and his disdain for Gottfried Leibniz ("he should have limited himself to a smaller sphere").[5] Wagner often tried to steer the conversation toward psychic phenomena, but Gauss was either too cagey or too indifferent to bite. At times he seemed put off by Wagner, at others politely interested, especially when the subject turned to the Romantic poets and novelists. What comes through most clearly is Gauss's lack of enthusiasm. He was a very sick man, and on several occasions he was too ill to receive Wagner. A week after their first meeting, he suffered a severe heart attack that almost killed him.

Carl Gauss did share many of Wagner's conservative leanings, but the two men were not close friends. Gauss was willing to admit his belief in vitalism, or at least in the possibility of a nonmaterial soul substance, as long as Wagner conceded that it would have to obey physical laws. It was not the ringing endorsement he was looking for, but this did not prevent Wagner from citing Gauss as a fellow traveler, and this infuriated Carl Vogt.

In 1855, the year of Gauss's death, Vogt responded to Wagner's antics with a book that is easily one of the most libelous rants in the history of science. *Blind Faith and Science* is a mockery of the career of Rudolf Wagner, a 150-page ad hominem attack in which he refers to Wagner as (among other things) an incompetent fraud, a plagiarist, a hypocrite, a showoff, and a "poison-swollen viper." But Vogt reserved his greatest contempt for Wagner's attempt to enlist Gauss in his fading cause.

"Whatever beliefs Gauss may have had I cannot say," Vogt wrote, "nor am I particularly interested to know. . . . When Newton got old, he wrote books about the apocalypse and burned his mathematical treatises. Gauss did not suffer the same sad fate that often comes with age, this softening of the brain."[4] Now that he was dead, however, the great mathematician could not speak up for himself. "Why didn't Wagner speak up earlier, when Gauss was still alive?" Vogt demanded to know.

According to Vogt's sources, Gauss was pained by the conversations with Wagner. Wagner was so excruciatingly polite and formal that Gauss found it difficult to disagree with him. "You yourself know what you do when you want to get rid of a pushy person," Vogt wrote. "You say politely in conversation 'yes, yes' to things you do not agree to, when if such behavior were allowed you would rather throw this colleague out the door."

Carl Gauss did not have to put up with Rudolf Wagner for very long. He died in his daybed at the Göttingen Observatory on February 23, 1855, two months after his final meeting with Wagner. His friend, Wolfgang Sartorius von Walterhausen, noted that Gauss's pocket watch, which he carefully wound every day, stopped ticking just moments after his heart did.[5] Konrad Fuchs, a professor of surgery, performed the autopsy in the presence of five observers, all Göttingen professors, including Wagner, who made the death mask and a cast of the inside of the skull cap. In an act that would resuscitate his career, Wagner took the initiative of preserving Gauss's brain. He did not have permission to do it, but he would get it later from Gauss's son, who thought that it might yield some interesting results.

He was right.

IT IS NOT clear how Wagner positioned himself to take possession of Gauss's brain, whether it had anything to do with the conversations. Wagner never mentioned a bequest, or suggested that his talks with Gauss addressed the issue of brain anatomy and the intellect. But acquiring the brain was a master stroke. It instantly legitimized a new line of research, one that would give Wagner the chance to live up to Blumenbach's legacy and to climb out of the shadow of his predecessor.

The man Wagner had replaced at Göttingen enjoyed an outsized reputation. Johann Friedrich Blumenbach is still remembered for his division of mankind into five races distinguished by color—white, brown, black, red, and yellow. As Steven Jay Gould discusses at length in an essay entitled "Three Centuries' Perspectives on Race and Racism," Blumenbach's most lasting legacy is the term *Caucasian*, which he employed to describe the white race in the mistaken belief that the European stock originated in the Caucasus Mountains of modern-day Armenia.[6] Blumenbach based his racial classifications on geography, on the fact that distinctive features predominate on each continent. Unlike many of his contemporaries (notably Georges Cuvier), he did not believe in a separate creation for each of the races, or even in a clear delineation of racial types. Instead he believed that all people are descended from a single pair, and that racial characteristics vary on a continuous scale. On aesthetic grounds alone he assumed that man's original stock was preserved among the inhabitants of a small region on the southern slope of the Caucasus Mountains. As these people drifted over the globe, anatomical variations began to appear in response to climate. Over hundreds of generations the features of the white race "degenerated"—literally diverged from the original stock—and the resulting adaptations, in Blumenbach's view, would be reversed over time if a racial group was transported to a different climate. In other words, the racial features that Blumenbach observed, including the skull shapes that were a central part of his data, were merely part of the ebb and flow of environmental influences on heredity.

Even though Blumenbach based his racial classifications on geographical rather than anatomical boundaries, the very act of creating and labeling racial categories implied the existence of an ideal type within each race. The same aesthetic grounds that led him to choose Caucasians as the original line of descent would lead him to propose a hierarchy of beauty in which Caucasians ranked first, followed by Native Americans and Malays (red and brown, respectively), with Orientals and Africans (yellow and black) degenerating furthest from the aesthetic ideal, and occupying the lowest rung on the ladder. As Gould points out, although Blumenbach's hierarchy was based solely in his personal idea of beauty, "ideas have consequences,"[7] and the consequence in this case was the birth of a scientifically sanctioned racism based on anatomical differences, in particular on the shape of the skull.

By the time Rudolf Wagner inherited Blumenbach's museum, with its famed craniological collection, improvements in techniques for preserving whole brains had allowed the focus of racialist studies to shift away from the skull and toward the brain itself. The race skulls were no longer of any use to Wagner. Gauss's brain, on the other hand, fascinated him. It was unlike any he had ever seen, and in order to quantify its uniqueness, he decided to acquire a control group. As it turned out, he was perfectly situated to do just that. And so, with few regrets, Wagner dropped out of his feud with Carl Vogt.

WHEN KONRAD FUCHS died at the age of fifty-two, less than six months after removing Gauss's brain, Rudolf Wagner was waiting. Before the decade was out, Wagner would collect three more brains of Göttingen professors—those of Carl Hermann, a philologist who died in 1855; of the celebrated number theorist Peter Gustav Lejeune Dirichlet in 1859; and of Johann Friedrich Hausmann, a mineralogist, also in 1859. He was not subtle about it, but even Gauss's brain could not open every door. For every brain he managed to acquire, another one got away, and he had to satisfy himself with data from autopsies. Even so, through sheer determination, Wagner managed to piece together a substantial collection, including brains not only of educated men, but also of townsfolk, notably a laborer named Krebs and a washerwoman, both of whom possessed remarkably large brains. The biggest brain of all belonged to a hydrocephalic man; the smallest to a microcephalic. Both were cases of mental impairment.

In addition to the brains themselves, Wagner collected brain weights. They came from an ever-expanding literature of postmortem examinations, beginning with Byron's. In all, he collected over nine hundred case studies, to which he added those of his five professors and of various townspeople, including the washerwoman and the laborer Krebs. The pathological cases filled out the high and low ends of the distribution.

In conducting his study, Wagner took Carl Vogt's criticisms to heart. Any work he did would be closely scrutinized, so he was determined to be thorough. Because he was not an expert on the human brain, he consulted the leading authority in the field, Louis-Pierre Gratiolet, who had published his atlas of the cerebral folds of primates in 1854. Wagner immersed himself in Gratiolet's diagrams and put together a brain study of his own, which he published in 1860 as "Preliminary Study on the Scientific Morphology and Physiology of the Human Brain as the Organ of the Soul." He could not resist the flourish in the title, even though Wagner's *Vorstudien* (its German title) is not really about the soul or the anatomy of genius. It is primarily a study of the comparative anatomy of primate brains, specifically the gross morphology of the cerebral cortex, including speculations on brain function based on anatomical structure at different stages of the brain's development.

But it was the centerpiece of the report—a list of 964 brains, ranged by weight from largest to smallest—that would overshadow everything else. It included the eight "elite" brains belonging to Byron, Dupuytren, Cuvier, and the five Göttingen professors. Tucked into a hundred pages of exposition and argument, the list spoke volumes.

SOPHISTICATED STATISTICAL TECHNIQUES are unnecessary whenever data forms a distinct pattern. If a clear relationship existed between the size or surface complexity of a brain and the quality of the intellect, it would have been obvious from Wagner's list. But no clear pattern emerged.

Patterns usually depend on the eye of the beholder, and often betray his or her preconceptions. To his credit, Wagner, who was hoping for a positive result, failed to see incontestable evidence: "The question of whether the number of folds is especially great in highly gifted or intellectually active individuals," he wrote, "cannot be deduced at this point."[8] He was equally cautious on the question of brain weight. "While certain brains of richly gifted men rank at the top of the list of brain weights," by which he meant Byron and Cuvier, "the fact that the brains of men of comparable significance, such as Gauss and Dupuytren, can be found in the second 100 . . . can only show the uncertainty of the earlier claim."

It was a laudable exercise of restraint, and also a quiet, perhaps intentional, rebuke of Carl Vogt. Wagner was a competent scientist. He would not let himself speculate beyond the data at hand. But the community of natural scientists was not happy about his caution. They thought he had misread the evidence.

The *Vorstudien* is aptly titled. It is a preliminary study, a promising start to the search for an anatomical explanation of genius and talent. No scientist would attempt anything of the same scope for another fifty years, including Wagner himself (he would not live to complete it). Its critics ignored its preliminary nature, however, and attacked it with an abandon that was directed more at its author than his conclusions. Carl Vogt, betraying his limitations as a thinker (or perhaps just his lingering animosity), commented that "Wagner of Göttingen has published a rather large table of cerebral weights, among which may be found those of many highly gifted persons; and he has based on this table the assertion that there is no good reason for the [brain-weight] theory, inasmuch as the brains of Hausmann and Tiedemann, who occupy an eminent position in science, were under the average weight. Exceptions, however, only prove the rule."[9]

He could not resist adding that "anthropologists with large heads may feel inclined to adopt one view, whilst those less favored will accept the opposite theory." Vogt, of course, had a big head.

RUDOLF WAGNER DIED in 1865, before he could complete his elite brain study. A second *Vorstudien*, published in 1862, dealt mostly with primate brains, although Wagner did investigate another possible marker for intelligence. If the gray matter gives rise to the intellect, he mused, then perhaps the mass of the cerebrum is less important than its surface area. Gauss's brain strongly hinted at this possibility. Its excessive fissurization suggested that its surface area was greater than that of brains of the same volume. To test the theory, Wagner embarked on the daunting task of measuring the surface area of a brain.

The process he devised is one that could only have occurred to a maniacal perfectionist. He cut small squares—4 mm by 4 mm—from "vegetal paper," and attempted to cover the entire cortex with them, both the outwardly visible parts, as well as the hidden surfaces within the folds. He managed to finish only a few brains before turning the project over to his son Hermann, also a trained physiologist, who had better luck using small squares of gold leaf. Hermann Wagner discovered that the extent of the cortical surface was far more extensive than previously thought. Two-thirds of it lies buried in the folds. But this new measure, surface area, used as a point of comparison between the brains of professors and peasants, proved to be as inconclusive as brain weight. When Wagner tried to calculate the relationship between the volume and the surface area, he discovered that even though large brains seem to be more richly convoluted than smaller ones, they have a smaller surface area relative to their size. With that, the Göttingen studies came to an end.

By 1865, Carl Vogt had also moved on. After reading Darwin's *On the Origin of Species* in 1859, he abandoned his insistence on the immutability of species (although not his belief in polygenism). In 1865, he published his *Lectures on Man,* which established his bona fides as an anthropologist. He continued to dabble in politics, but he devoted most of his energy to marine biology. The angry young man, no longer young or angry or driven to change the world, went back to doing what he enjoyed most, biological research.

In the decades following its removal, Gauss's brain would be used by both sides of the materialism conflict. In Paris, where the search for the anatomy of genius relocated in the 1860s, it would serve as the centerpiece in a heated debate over the localization of brain function and the existence of the soul. It would also inspire the founding of several brain donation societies. But rather than resolving questions, it only seemed to multiply them. Today Gauss's brain sits in a cabinet in the basement of the Institute for Medical Ethics and the History of Medicine on the campus of the University of Göttingen, where it keeps company with the brains of Dirichlet, Fuchs, and Hermann. The fifth member of the original group, the brain of Hausmann, has been lost. After Hermann Wagner laid them aside in 1865, the brains were left to gather dust at the Department of Anatomy and Physiology at the university. There they remained, undisturbed, until 1970, when a professor of pathology, Hans Orthner, decided to take charge of them. For twenty-five years, Orthner kept Gauss's brain on his desk, and in 1995, well beyond retirement age, and sensing the weight of responsibility, he offered the brains to an up-and-coming historian of the sciences, who delivered them to the institute.

Not long afterward, Gauss's brain was removed from its sealed jar and subjected to magnetic resonance imaging. It revealed no sign that Gauss had suffered mental degradation in his final years. Its excellent shape testified to Rudolf Wagner's skill as a technician. Although the brain scans have been available for some time to anyone interested in studying them, no one has seized the opportunity. No one, for example, has bothered to compare Gauss's brain to Einstein's, for reasons that have nothing to do with science. In the lottery of fame and fortune, which is to say, in the eyes of popular culture, Gauss's intricately fissured brain is simply old news; and so, in the hype-driven lottery of genius, is Gauss himself.

Chapter 6

Broca

THE MUSÉE DE L'HOMME, or Museum of Man, occupies one of two gracefully curving pavilions that make up the Palais de Chaillot, an Art Deco acropolis that rises above a bend in the river Seine. From its upper floors, one can look across to the Eiffel Tower to the southeast, and beyond to the Champs de Mars. The view to the northwest, if not quite so dramatic, is at least pleasant, taking in the Place du Trocadéro, the Cimetière de Passy (where Édouard Manet and Claude Debussy are buried), and a statue of Benjamin Franklin that sits on a small traffic island known as Yorktown Square. This is the view that greets visitors to one of the museum's hidden treasures, a collection of moldering anatomical specimens that commands a sunlit space on the third floor, far from the bustle of the visitors' galleries.

Recently, the museum's administrators decided that this prime piece of real estate was being wasted on useless junk. In their view even the basement was too good for an odd lot of skulls, bones, busts, and brains that had accumulated on the premises in row upon row of tall metal cabinets, and they decided to ship all of it to a remote site. They were forced to rethink this plan, however, when it was pointed out that the "junk" in question represented a significant chunk of Paris's past, if not an important chapter in the history of science.

One of the cabinets contains, in its entirety, Franz Josef Gall's phrenological collection, essentially the raw data behind his doctrine of the skull. It includes some of the most celebrated heads in Paris—a bust of Goethe, plaster masks of Franz Liszt, Voltaire and Rousseau, even a cast of Descartes's skull (the original is kept in another part of the museum), and three phrenological busts of Gall himself. But the real prize, in an adjacent aisle, is the set of brains that inspired Carl Sagan's celebrated essay "Broca's Brain" (which happens to be the least significant thing about them). What justifies their place next to Gall's busts is their role in perpetuating his phrenological program. The brains in question are not so important for *who* they were as for what their owners did. These were the men who founded the world's first brain-donation society. They also represented the French response to Rudolf Wagner's study of Gauss's brain. Which is to say, these brains were extracted and preserved in order to make the case for the importance of brain weight, for the significance of fissural patterns, and for the localization of brain function. What the museum directors eventually came to realize was that in these two cabinets they had accidentally preserved the forensic evidence that finally tipped the scales in the battle against the soul initiated by Carl Vogt.

BROCA'S BRAIN IS no longer among the brains kept at the Musée de l'Homme. It never really belonged there, and in fact it played a far less significant role in this story than Broca himself.

Paul Broca lived in Paris from 1840 until 1880, first as a student, then as a surgical resident, eventually as a full professor. In the final year of his life, he was even elected a senator. During a prolific career, he invented tools and techniques in neurosurgery, he was the first to describe the pathology of muscular dystrophy, he pioneered the practice of blood transfusions, and he almost singlehandedly established physical anthropology as an academic discipline. He was a crusader for social justice, and promoted universal education and health care. He was also a groundbreaking anatomist, as a glance at any anatomy text will show—Broca's area, Broca's convolution, Broca's band, and the limbic lobe—these and many other structures were named for or by him. Over an impressive career that combined experimental work with thousands of clinical observations, Broca authored over five hundred medical and anthropological papers, and was by the time of his death ranked with the likes of Louis Pasteur.

Like all lists of accomplishments, this one ignores the many daunting obstacles that stood in the way of the discoveries—not just the lack of technical means, but the hostility of a status quo threatened by any form of change. In this case, the threat was very real. Broca's work as a surgeon was complicated by the fact that he was also a humanist, which is to say that he hoped to establish the study of human beings as a part of the natural world instead of the spiritual one. His goal was not to eliminate spirituality from public life, but from the way he went about his business, this was not obvious, given that one of his first orders of business was to found an anthropological society in Paris. He made it happen in 1859, and it wasn't easy.

During most of the previous year, while Rudolf Wagner carefully assembled his small collection of special brains in Göttingen, Broca was busy calling in favors from his many friends, and twisting enough arms to get eighteen signatures on a petition for a controversial scientific society. In the process he had been forced to run a bureaucratic gauntlet while being passed off between the minister of public education and the prefect of police, each of whom thought the young man would run out of patience. But they were wrong. Six months after he first applied for a license, Broca's Société d'Anthropologie de Paris met for the first time.

Broca succeeded through the expedient of keeping the membership below twenty, the legal limit for gatherings. He also agreed to keep politics and religion off the table. To ensure that he did, the prefect assigned a policeman to attend every meeting as an official eavesdropper. Within two years the surveillance was lifted, and the Société was officially recognized. Within five years it was accorded the full status of a "public utility."[1]

How this could have occurred under any sort of vigilant surveillance is unclear. According to Sagan, whenever the policeman asked permission to take a break, Broca would insist that he sit and earn his pay. Had he been attentive, the prefect's spy would have heard discussions that came closer to true subversion than even a casual listener could miss, for in the second year of the Société's existence, Rudolf Wagner's *Vorstudien* was placed into evidence, and the question of the soul came into the open. Most of Broca's colleagues, and Broca himself, were inclined to doubt its existence.

The Société d'Anthropologie belonged to Paul Broca in the sense that he directed its agenda. He had declined the presidency, knowing it would limit his participation while calling too much attention to himself. He insisted on keeping the minutes instead, and continued to do so for the next fifteen years, effectively controlling the agenda of the Société from the secretary's chair.

During the first year, the topics raised at the Société's meetings ranged over the cultural heritage of France and the ethnic origins of the French people. But the subject that worried the prefect of police crept into the agenda in January of 1861, when Louis-Pierre Gratiolet, one of Broca's eighteen "co-founders," brought in an unusual skull. He had been waiting for this opportunity for some time. The specimen gave him a pretext for broaching the most vexing anthropological issue of the day—the relation of brain anatomy to intelligence.[2]

The skull belonged to a Totonaque, a member of a Mexican coastal tribe. It was exotic enough, from a Parisian's perspective, to be news at the Société. Anything new (and in the world of physical anthropology just about everything was new) was exotic, be it the skull of a Totonaque, the head of a Fijian, or the brain of a gorilla. The anthropologists acquired such specimens from sea captains, from explorers, from zoos, and from traveling circuses. Their precious samples, many of them one of a kind, had the dubious honor of representing entire species, if not races. What intrigued Gratiolet about this skull was its size. It was far bigger than the average Parisian's skull. While it was not his only piece of evidence, it seemed to verify Rudolf Wagner's contention that brain size was not a reliable measure of intelligence. Gauss's brain was another case in point. It was only slightly above average in weight, but possessed an intricate pattern of convolutions and fissures. In Gratiolet's view, the evidence seemed to confirm that brains are "only as perfect as the convolutions are developed."[3]

At that time, especially in Paris, this was not simply a scientific opinion. Ever since the rift between Franz Josef Gall and Pierre Flourens, the scientific community had been divided over phrenology, specifically over the localization of mental faculties. At the center of this conflict was the existence of the soul. Gratiolet, like Wagner, held to the party line: just as the soul is one, the cerebrum must function as a unit; and because of the presence of an immaterial vital force, the size of the brain (within certain limits) was irrelevant.

As a freethinker, Broca took the opposite position. Like his fellow republican anticlericalists, his goal was to establish an independent footing for science without the interference of church or state. By proposing mankind as a suitable object for objective, empirical investigation, he hoped to eliminate the distinction between humans and the natural world, a distinction that had thus far been artificially propped up through the expedient of the soul. Although Broca was probably the most radical among the French freethinkers, he was far more subtle than Carl Vogt, and more politically astute. Through his anthropological society he was starting a quiet revolution, and his good friend Pierre Gratiolet would have the privilege of serving, unknowingly, as a useful foil.

PAUL BROCA AND Pierre Gratiolet were born ten years apart in the small Burgundian village of Sainte-Foy-la-Grande, a Huguenot stronghold near Cyrano's Bergerac. Their fathers were physicians, but the Gratiolets had the misfortune of being Catholic in a tightly knit Protestant community. In 1826, when Pierre was ten years old, his father, who had been unable to build a clientele among his suspicious neighbors, relocated the family to Bordeaux, where he did not fare much better.

From early childhood, it seems, Gratiolet was a hard-luck character of Dickensian dimensions. He was a bright student, and he set off for Paris at the first opportunity. Once there he studied medicine and earned a degree, although he never practiced as a physician. He gave up the idea the instant he set foot inside the Museum of Natural History, the house that Georges Cuvier had built. In that hall of wonders Gratiolet stood transfixed, not just by the sheer variety of nature, but by the uncanny similarities between species. He decided to study comparative anatomy with Henri de Blainville, a devout Catholic and royalist who had taken over the museum when Cuvier died in 1832. Perhaps, Gratiolet thought, he would eventually inherit Cuvier's chair himself. And so he toiled away in Blainville's laboratory, assisted François Leuret in producing the first volume of his atlas of primate neuroanatomy, lectured, wrote, and distinguished himself in a branch of anatomy that few men had ventured into. When Blainville died in 1845, Gratiolet stood at the very threshold of greatness. By rights, he should have succeeded to Cuvier's chair. But he was passed over. Despite his outstanding credentials, his groundbreaking research, and his popularity, Pierre Gratiolet would remain an impoverished laboratory assistant at the museum for most of his professional life.

By contrast, Broca, who was ten years Gratiolet's junior (they never did meet as children), enjoyed one success after another. The Brocas were Calvinists, and they thrived in Sainte-Foy. Paul excelled at school, and upon reaching Paris he became, at age seventeen, the youngest prosector ever at the Société Anatomique. From there he would progress rapidly from surgical resident to full professor at the École de Medicine. He did not accomplish this by toeing the line. Broca would be singled out as the most radical voice in an increasingly politicized faculty, and also as the most talented surgeon. Had he arrived in Paris ten years earlier, when Gratiolet did, his outspokenness would probably have limited his prospects. But the climate was changing rapidly, and freethinkers like Broca quickly learned how far they could press their luck.

It is not clear whether Broca's religion grew out of his politics, his politics out of his religion, or whether, as seems most likely, his most fundamental beliefs grew out of his passion for science. Although he was a Protestant, and married a Protestant in a religious ceremony, he once said, probably arguing from his own experience, that "human beings become less religious as they become more educated." He regarded religion as "nothing more than a submission to authority."[4]

Pierre Gratiolet, on the other hand, was a devout Catholic. Education had not quelled his religious fervor. Almost single-handedly, he argued the conservative viewpoint at the anthropological society, which is to say that he pressed the case against phrenology. "In a general way," he said, "I agree with Monsieur Flourens that intelligence is one, that the brain is one, that as an organ it acts as a whole."[5]

Like most of his colleagues, Gratiolet believed in the superiority of the European race, and he sought confirmation of it in his anatomical studies. One of his more egregious errors, announced at one of the first meetings of the Société and brandished by its members for the next twenty years, concerned the sutures of the skull, the fault lines that allow it to expand to accommodate the growing brain during childhood and adolescence. Gratiolet believed that in the "less perfectible races" the sutures close sooner than in the white race. Moreover, the sutures in the skulls of blacks close from front to back, whereas the reverse is true in whites. As a result, the overall growth of the brain is restricted in inferior races, especially in the all-important frontal lobes. (This conclusion, based on the examination of only a few skulls, was soon shown to be false.)

In arguing for the unity of brain function, Gratiolet built his case on analogy rather than evidence. His most compelling argument was that no one could "attend" to two things at once. If the brain was made up of modules, he said, it would stand to reason that it could process several lines of thought simultaneously. But this is clearly not the case.

Instead, what did seem to be the case, as an increasing number of clinical studies began to show, was that damage to specific parts of the brain could cause the loss of very specific skills. Many such cases were being reported at meetings of the Société. If the brain did operate as a whole, as Gratiolet claimed, how was it possible for a highly localized lesion to destroy something as specific as the ability to associate a name with a face?

APHASIA IS THE technical term used to designate impairment of the ability to express ideas or symbols in words. Initially, Paul Broca used another word—*aphemia*—to describe a condition that afflicted his patients who were, as he put it, missing "only the faculty to articulate words." But Broca's coinage never caught on, and aphasia (ultimately "Broca's aphasia") took its place in the literature and remains in use today.

In the summer of 1861, aphasia came to play a central role in the debates at Broca's anthropological society, largely because speech was regarded as a uniquely human faculty: that is, the ability to use and understand written and spoken language, it was thought, set men apart from animals. Descartes cited this in his argument for the dual nature of mind and brain, and it continued to be regarded as the best evidence for the existence of the soul. But if the brain could be shown to possess a special module dedicated to producing words, then the key difference between man and beast would be reduced to nothing more than an anatomical peculiarity in man's brain, and man's very uniqueness would be in jeopardy.

Cases of speech loss had been described in great detail long before Broca gave the syndrome a name. Franz Josef Gall, among others, had recorded several interesting case studies. Among these was a young man who had suffered a fencing wound that pierced "the middle part of the left canine region, near the nostril." From there the point of the foil had passed "in a vertical direction and a little oblique from before backward, to the depth of five or six lines in the internal posterior part of the anterior lobe of the brain in such a manner to approach the anterior part of the mesolobe."[6] (The strained wording of this passage illustrates the awkward state of affairs before Gratiolet named the parts of the cerebrum.) Although the patient could still speak, he had lost almost all memory of names, and began to refer in most instances to "Monsieur Such-and-Such." To Gall, this seemed to confirm the existence of a faculty of verbal memory, and of a brain module dedicated to "attending to and distinguishing words." It was the very faculty of mind that he had correlated with bulging eyes, and had placed in the frontal lobes.

Similar cases of speech loss came to light in the 1820s, but Gall did not exploit them. Instead, one of his students, Jean-Baptiste Bouillaud, carried on this work, insisting that the only way to go about finding the speech center was to correlate symptoms with brain lesions. After amassing hundreds of case studies, Bouillaud was satisfied that "the principal lawgiver of speech is to be found in the anterior lobes of the brain."

At about the same time, an obscure physician from Montpellier, Marc Dax, was conducting similar studies, and, in 1836, he presented a paper to a regional medical society in which he argued for the existence of a speech center not just in the frontal region, as Bouillaud and Gall believed, but specifically in the *left* frontal lobe, a claim that contradicted Gall's belief in the symmetry of brain function. The idea was too radical for its time, and Dax's paper would languish in obscurity for thirty years, while Bouillaud's work was so tainted by its association with phrenology that it too failed to win support.[7] So frustrated did Bouillaud become that in 1848, after pleading his case to the Faculté de Médecine, he offered a bounty of five hundred francs to anyone who could produce an instance of a patient with a severe lesion in the frontal lobes who did *not* suffer from a speech impairment. There were no takers, but the absence of contradictory evidence failed to establish Bouillaud's claim, and the issue went unresolved for the next twelve years, until Pierre Gratiolet made his presentation of the Totonaque skull and argued against the localization of speech. Paul Broca took the opposite position, and sided with Bouillaud.

In his own work, Broca had seen several cases of speech loss accompanied by lesions of the frontal lobes, yet he did not make this the centerpiece of his case against Gratiolet. In a series of debates that dragged on through the summer of 1861, Broca portrayed Gratiolet's belief in the insignificance of brain weight and the unity of brain function as a threat to the legitimacy of anthropology. If his learned colleague was right, Broca warned, "then the study of brains of human races loses the greater part of its interest and utility." This was beside the point, of course, but Broca was not about to let Gratiolet win.

In subsequent sessions, Broca would argue, rather archly, that if the hydrocephalic brains were discarded from Rudolf Wagner's list of 964 brain weights, then Lord Byron and Georges Cuvier would rank first and second. "This seems more than coincidence," he observed.[8] As a matter of national pride, Broca was unimpressed by Wagner's dead colleagues, and he refused to accept them as examples of great intellects in small heads. "A professor's robe is not necessarily a certificate of genius," he said. "There may be, even at Göttingen, some chairs occupied by unremarkable men."

It is the irreverent and domineering Broca of these exchanges who rightly earns Steven Jay Gould's mistrust in *The Mismeasure of Man*. When arguing against the evidence, Broca could be especially mean-spirited, and often simply uninformed. "The question is whether Gauss and Dirichlet were as great among mathematicians as Georges Cuvier among naturalists," he remarked, before concluding, somewhat naively, "I don't think so. Cuvier manifested more aptitudes than these two men to whom he is compared."

Gould was particularly concerned about Broca's racial views, specifically his brain-anatomical arguments for the superiority of white Europeans. If there is any defense for such opinions, unsupported as they were by a reasonable statistical analysis, it is that Broca's goals were not social, or even scientific, but political. In buying into the materialist position, Broca had accepted polygenism—the separate creation and lines of descent of each of the races. According to this theory, some races had advanced further than others, as evidenced by their technological accomplishments. Broca assumed that this state of affairs was largely a product of environment. The inferiority of women, for example—specifically their undeniably inferior intelligence, their smaller and less-developed brains—had resulted from their subservient role in society. Presumably, with increased stimuli, their brains would develop within generations, and Broca used this point to argue for the reform of girls' schools. "Not only does education," he wrote, "make a man superior . . . It even has the wonderful power of raising him above himself, or enlarging his brain, and of perfecting its shape. To ask that instruction ought to be given to everyone is to make a legitimate request in the social and national interest, and perhaps in the racial interest beyond that. Spread education, and you improve the race."[9]

Unlike many of his fellow polygenists, Broca never argued for the subjugation of any one group, or for the withholding of rights. In the matter of race, according to his biographer, Francis Schiller, Broca "admitted differences in intelligence, aesthetic appeal, or military power," but he also "found the [monogenist] argument of a physical or moral degradation, used in order to maintain slavery, repulsive."[10]

In his debates with Gratiolet, Broca chose to play the role of skilled politician more often than that of objective scientist. His paramount goal was to free academic inquiry from restrictions imposed by the church. In order to do this, he had to adopt the materialist (essentially the phrenological) position, and follow it to what seemed to be its logical conclusions. In retrospect, as Gould showed, Broca's selective interpretation of his data was not logical at all. Even so, he did manage to get one big thing right.

IN MOST CAPSULE biographies, Broca's legacy has been boiled down to a few signal accomplishments: his invention of physical anthropology, his use of the tools of that trade to establish a hierarchy of races and the inferiority of women, and finally, in anatomy, his vindication of the principle of cerebral localization through his "discovery" of the module of speech, now known as Broca's area. There is more to these accomplishments than meets the eye, but the last of them is particularly interesting. In what sense did Broca prove the localization of speech?

The key incident, the supposed Eureka moment in the history of neuroscience, occurred in April of 1861, in the midst of Broca's rebuttal of Gratiolet. Broca came to the meeting that day with an unusual brain specimen. He did not intend to enter it into evidence against his opponent, nor did it stop Gratiolet dead in his tracks. Although many historians have portrayed this as Broca's Perry Mason moment, the rather mundane fact of the matter is that Broca was merely registering the specimen before depositing it at the Musée Dupuytren.[11]

The case was an interesting one. It involved a man who had resided at the Bicêtre hospital for the previous twenty-one years. During that time he had suffered from a progressive speech loss and a paralysis of the right side. He stood out from Broca's other patients because he could utter only one sound, which he did repeatedly, using it to convey the full range of ideas in his head. His name was Leborgne, but Broca chose to refer to him by this strained utterance: "Tan." Leborgne could understand everything said to him, which meant that he hadn't lost his understanding of words, merely the ability to produce them. Oddly enough, in addition to the word "tan," he retained a single epithet, which he used whenever he failed to make himself understood. "Sacré nom de Dieu!"—Goddamn!—was Tan's only available profanity, and he used it frequently. When Tan died on April 17, Broca performed the autopsy and found that a lesion the size of a hen's egg had destroyed the third frontal convolution (located somewhat above the temple) on the left side of Tan's brain. Significantly, the right side was intact.

The phrenological school, beginning with Gall, took for granted the symmetry of brain function, basing this notion on its symmetry of structure. Although no two hemispheres of any brain are perfect mirror images, just as no two sides of a single face are quite alike, they are similar enough to have assured Gall that each of his psychological faculties must be duplicated in each hemisphere. It seemed too improbable that the left and right sides of the brain would divide the labor. The duplication of the eyes and ears provided another compelling analogy. One researcher of the 1840s had ridiculed the idea of functional asymmetry, saying that it was "as though the loss of one eye would cause blindness."[12] Saddled with the same prejudice, Broca hesitated at first to conclude that speech was a left-brain faculty. He had seen other cases of speech loss involving lesions of the left frontal lobe, but he had also come across exceptions to that rule. In his presentation of Tan's brain to the Société, he decided not to allude to the significance of the left side at all.

As Pierre Marie, the celebrated clinical neurologist (and one of Broca's students), would point out years later, the brain of Tan was a poor choice to represent the argument for a speech area.[13] For one thing, his lesion was not limited to the third frontal convolution. It was merely centered there, and it had destroyed some of the surrounding tissue. Because Broca refused to dissect the brain, the extent of the lesion could not be determined for certain. There was, however, an even larger problem which did not escape Broca's analytical mind. The phenomenon of speech is a very complex and multifaceted process that draws upon memory, imagination, and motor skills, among other components. These components are not only difficult to identify, but sometimes impossible to distinguish from each other, or even define. So it is not surprising that Broca hesitated to name the third left frontal convolution as the seat of the mechanical production of speech based solely on the case of Tan. When a similar case showed up six months later (this time a man named Lelong, who had a smaller lesion contained entirely *within* the third left frontal convolution), Broca presented the brain to his colleagues, again without proposing speech as a left-brain function. (He would not make this case until three years later.) Although these two presentations have since been reinterpreted as momentous events in the history of science, their significance went unnoticed at the time. In fact, after the presentation of Tan's brain, which would seem to have demolished his case, Gratiolet returned to the podium and calmly resumed his argument.

THROUGHOUT THE 1800S, the ideological split over phrenology continued to divide medical men into two camps. There seemed to be no middle ground. As far as Pierre Gratiolet was concerned, Franz Josef Gall was guilty of "the most ridiculous analysis ever made about the faculties of human understanding."[14] Broca, on the other hand, along with a growing number of scientists, believed that Gall was correct in theory, even if he was woefully wrong about the details. The brain was a system of components, they said, and Gall had merely labeled them too capriciously.

"Consider the intellectual functions," Broca said, "how they frustrate all attempts at identification. Some are called 'faculties,' others 'qualities,' still others 'sentiments,' 'penchants,' or 'passions.' There is no generic name that encompasses them all."[15] Broca was not about to repeat Gall's mistake and try to name the brain's divisions. "The functions of the brain," he said, "are simply too diverse, and reveal themselves in diverse ways. . . . For example, there are idiots who have extraordinary memories; there are illnesses that leave judgment intact while destroying memory; and just to show that memory is not a simple faculty, there are lesions, even purely traumatic lesions, that abolish only a part of memory, like, for example, the memory of proper names or of facts of a certain type or facts from a certain time."

Gratiolet preferred to lump all brain functions together. "Some psychologists have made a great error," he said. "They have decided that sensation, imagination, and memory are distinct faculties in the soul, whereas in reality they are nothing but the modes of a single primitive faculty, the faculty of knowing."[16] But knowing *what*?

The two men agreed about this much: The human intellect is a product of the human brain, and the thinking part of the brain resides in the cortex. They also seemed to agree, as Broca put it, that "the pondering of intelligence should precede the pondering of the brain." But instead of trying to define intelligence, they argued over it. Broca, for example, tried to distinguish between general intelligence and a more limited, subject-specific intelligence, which led him to make a dubious claim. He insisted that Georges Cuvier had been much more accomplished in his sphere than Carl Gauss. Gratiolet testily countered that, "I have seen, in effect, among men with large heads, as well as among those with small heads, men of great spirit, mediocre men, and fools."[17]

At least one of the anthropologists, pondering the implications of this, saw no way out of the dilemma. His name was Eugène Dally, and he represented the future of the Société. Just twenty-eight years old, and only two years out of medical school, he had been reluctant to speak, but he now took the floor. "The expressions 'intellectual superiority' and 'degree of intelligence,' " he said, "strike me by their lack of precision. The question, already complex enough when applied to men compared to each other, becomes in a way insolvable if you extend the comparison from animals to man."

The rest of Dally's speech deserves quotation in full:

> When we express a judgment on the relative value of two men, saying that one of the two is more intelligent than the other, it is an individual assessment, excessively variable, and on the subject of which a wealth of accidental elements play an important role; in a word, it is a judgment that has no scientific character. Education, the social milieu, hygiene, and even luck, all fortuitous circumstances, inspire different assessments of the apparent value of men and also affect the exactitude of opinion that we form of their merit. What is more, it is doubtful that there exists a common measure, all other conditions being equal, between a Leibniz and a Bacon, between a Cuvier and a Geoffroy Saint-Hillaire, between a Racine and a Corneille. I see everywhere diverse aptitudes, qualities in some way specific; these aptitudes and qualities, I find them if I extend the comparison to ethnic groups. But I search in vain for a positive criterion which permits me to affirm, without preconceptions, the superiority of this intelligence, of this nation, or of this race. I beg to ask my colleagues to well determine the positive signs with which one can recognize, among men, individual superiority and ethnic superiority.[18]

He had a point, but Dally's objection fell beyond the immediate concerns of the anthropologists. If assessments of relative worth were to be taken off the table, as Broca had said at the outset, the entire discussion would have lost all interest. As Stephen Jay Gould has pointed out, the quest to quantify intelligence "as a single number capable of ranking all people on a linear scale of intrinsic and unalterable mental worth" would continue unabated down to the present in various guises, none of which has withstood close scrutiny. It is disappointing, but not surprising, that Dally's point was not raised again.

Before the year 1861 was out, Broca and his supporters (essentially the combined membership of the Société) would wear Gratiolet down, and force him into a concession speech in which he did not abjure his original argument, but insisted that his audience had not appreciated its subtleties. But subtleties were not the point. Gratiolet's position was, in the eyes of its detractors, essentially political, and a threat to the progress of science. In that sense, the debate was not really about the brain or about intelligence, but about who had dominion over them.

BY THE 1870S, the Société d'Anthropologie de Paris had attracted hundreds of new members, but the meetings themselves were dominated by Broca and a core group of freethinkers who disavowed allegiance to any church. If they believed in God at all, they kept quiet about it. At the same time, the Société was also attracting a growing number of militant atheists, for whom scientific materialism had become a form of fundamentalism in itself. To these men, even a moderate doctrine such as the one advocated by Auguste Comte posed a threat to their agenda. Church burials in particular had become a sore point, if not *the* sore point. Some of the atheists went so far as to declare in their wills a desire to be cremated, a practice that was expressly forbidden by the Church of Rome. Others insisted on autopsies, either for scientific purposes (one donor wrote, "I will furnish, when the time comes, the most beautiful case of ataxia one could ever hope to see"), or simply for spite. ("I do not want," wrote another, "after my death, to contribute, even a little, to the accumulation of the wealth of the clergy.")[19] Another common reason for insisting on a postmortem examination was the prevalent fear of being buried alive.

If one goal united the positivists and the materialists, it was their effort to demystify and desanctify the corpse. Their principal tool for doing this was a showroom of human body parts, the Museum of Anthropology (later called the Broca Museum), which Broca had started with his personal collection of twenty-five hundred skulls, to which the society's fifteen hundred skulls and its other holdings—preserved heads, bones, casts of brains, articulated skeletons, and the world's largest collection of human hair—had been added. To get to their meeting place, the members had to pass through the dissection and preparation rooms, with their various works in progress, and then through the museum itself. In their enthusiasm for anatomical investigations, some of the more radical atheists declared that, rather than be buried, they would rather become part of the exhibit. This was how the first autopsy society got its start.

ACCORDING TO THE *Bulletin* of the Société d'Anthropologie, in October of 1876 Auguste Coudereau, a physician, proposed to found a group within the Société "for the purpose of increasing the number of autopsies and thereby enriching science by compiling anatomical documents that will place deductions on firmer ground."[20] Those who signed on immediately were Louis Asseline, a jurist; Yves Guyot, an economist; Louis-Adolphe Bertillon, a demographer; Henri Thulié, a physician; and four accomplished natural scientists: Charles Letourneau, Abel Hovelacque, Eugène Véron, and Gabriel de Mortillet. Paul Broca did not join. Presumably, he saw the group as political theater, an attempt to shock bourgeois sensibilities, which is precisely what it was.

The climate in Paris had changed considerably in the sixteen years since the Broca-Gratiolet debate. Gratiolet had died in 1865, just two years after attaining the professor's chair that had eluded him for twenty years. (Broca gave the eulogy, in which he was careful to make no mention of Gratiolet's most heartfelt belief—that he possessed a mortal soul.) During the 1860s, several key players in French science had returned from exile, notably Gabriel de Mortillet, who can best be described as the second coming of Carl Vogt.

The two men's biographies are remarkably similar. Both de Mortillet and Vogt were exiled for their roles in the Revolution of 1848. Like Vogt, de Mortillet went to Switzerland, and for a while worked as a zoologist in museums in Geneva and Annency (where he met and befriended Vogt). He was eventually expelled because of his political views. From Switzerland, he crossed over into Italy, where he served as an engineer on the construction of rail lines. There he indulged a passion for archaeology by exploring the lakes of Lombardy, and he discovered a Neolithic settlement on Lake Varese. After returning to France in 1864, de Mortillet joined Broca's Société d'Anthropologie with the intent of tearing down the legacy of Georges Cuvier, in particular Cuvier's belief in the immutability of species. He chose prehistory as his specialty, and became widely known for his classification of prehistoric epochs. He also befriended Broca, and became one of his closest allies within the Société. But unlike his good friend, de Mortillet was a committed atheist, committed specifically to the elimination of religion from civic institutions. With the advent of a republican government in 1870, he would finally get the chance to put his ideas into action. As the mayor of Saint-Germain from 1882 to 1885, he tried to secularize his district through every conceivable measure, going so far as to order the renaming of streets (he wanted rue Saint Louis changed to rue Diderot), and the removal of the cross from the local cemetery. Most of these changes would be reversed by his successor.[21]

In 1876, the year the autopsy society was formed, Broca capped off a long struggle to found a school of anthropology by appointing a faculty that included de Mortillet as professor of prehistory. Paul Topinard, Broca's protégé, would teach biological anthropology, Eugène Dally would teach ethnology, Abel Hovelacque linguistics, Louis-Adolphe Bertillon demography, Arthur-Alexandre Bordier medical geography, and Broca himself physical anthropology. Three of the seven men—de Mortillet, Bertillon, and Hovelacque—were radical materialists and highly astute propagandists. (De Mortillet, like Vogt, was a successful popular science writer.) Their goal was to spread their gospel from the pulpit of the Société d'Anthropologie and the School of Anthropology, which they now set their sights on controlling. Broca did not see this coming, even though his friends had already founded a Masonic Lodge of Scientific Materialists and a Scientific Materialist Dining Club. Their next move was the founding of an autopsy society.

The Société Mutuelle d'Autopsie was open to anyone willing to sign a donation form and fill out a questionnaire. Its founders promoted it in popular journals, and the response was enthusiastic. Prospective members had the option of signing either a standard bequest form, or one with an optional clause for freethinkers that read: "I . . . demand, as a logical consequence of my convictions, that the burial of the parts of my body that the laboratory does not keep for its studies will be done without any religious ritual and that the ceremony be purely civil."[22]

Even this did not go far enough for some members, who composed their own pledges. One of these, seemingly modeled on a fraternal blood oath, stated: "Free thinker, loyal to scientific materialism and the radical Republic, I intend to die without the interference of any priest or church. I bequeath to the School of Anthropology my head, face, skull, and brain, and more if it is necessary. What remains of me will be incinerated."[23]

The first member to honor his pledge—in 1878—was Louis Asseline, a writer and radical activist. Broca performed the autopsy. Although he found nothing distinctive about Asseline's brain, he diplomatically noted, "It is interesting for the single fact that it belonged to a man of remarkable intelligence." Its weight was well above the mean, but it was not a delicate brain. Broca described its convolutions as thick, almost coarse. Moreover, he found a fissure, the so-called ape fissure, which Gratiolet had regarded as a sign of inferiority. "This fissure is often prominent in women, thus indicating in some men an intellectual mediocrity," Broca said. But he quickly added that Asseline "was of an intelligence that could not have been more distinguished, and of a remarkable erudition."[24] The alleged sign of inferiority, he concluded, needed further investigation.

Four years later, Asseline's brain was formally critiqued before the membership of the Société d'Anthropologie. Mathias Duval, presiding from the chair, proudly announced that "the report we present today is, at least in France, the first report in the archives of this genre, archives that we hope will constitute an anthropological laboratory, thanks to the work of the Mutual Autopsy Society." The self-congratulation was premature. A few of the members were offended by the glib self-righteousness of the materialists and their disrespect for the deceased. One of them, a retired naval officer named Foley, was particularly incensed by the reference, repeated by Duval, to the ape fissure in Asseline's brain. "Owing to the communication of Monsieur Duval there is no need for further argument against such a society and its immorality," Foley grumbled. "Monsieur Duval himself said that in a way, the brain of Asseline had simian features. I refuse to pay any compliment to a society that makes such announcements. You have not made a very happy debut."[25]

A quick rejoinder by Paul Topinard quelled the dispute. "Monsieur Foley has chosen a phrase that is out of place. Happy? in what sense? He does not assume that we are pursuing a goal, that we hope to support a cause. We seek the truth, and nothing else. . . . The Société d'Anthropologie does not belong to any sect." But to his own chagrin, Topinard would soon learn otherwise. In the 1890s, the radical materialists, who had by then captured the key positions in the School of

Anthropology, expelled Topinard by changing the lock on his office door. The coup was then complete.

As the brains began to accumulate, it became clear that the Société's directors did not know what to do with them. Their first attempts at assessing their value, along with the ruckus over Asseline's brain, convinced them to limit their reports to exacting descriptions of the brains, rather than qualitative assessments. But even this caused problems. The most important brain in their possession, that of the statesman Leon Gambetta, was by far one of the smallest brains they had ever seen, so small that they conveniently omitted the brain weight—a mere 1,160 grams—from the autopsy report. Gambetta had been the leading light of French republicanism. A brilliant orator, he had risen to the presidency of the Senate, and then, while at the height of his power, had succumbed to appendicitis at the age of forty-four. The anthropologists debated what to do about this small brain, and they were forced to concede that brain weight, while important, must not be all-important.

For a group that hailed the objectivity of science as their new god, the materialists were surprisingly sentimental when it came to the postmortem examinations of their friends. When one of the founders, Louis-Adolphe Bertillon, died in 1883, his brain became the subject of a comparison with that of Gambetta. The two men were opposites in one glaring respect. Gambetta, a brilliant speaker, dazzled audiences with his erudition, while Bertillon struggled to make himself understood. According to the principal investigator, a M. Chudzinski, poor Bertillon "expressed himself with difficulty, searching for words and painfully constructing his sentences."[26] Yet nothing remarkable in Broca's region of either brain would confirm such a diagnosis. Charles Letourneau, another of the radical materialists, tried to rationalize this finding. Despite his difficulties expressing himself, he said, somewhat imaginatively, Bertillon was an orator at heart with a knack for imaginative, poetic, and metaphoric language—in other words, a "psychic orator," a great intellect betrayed by a faulty means of expression. "This imperfection," said Letourneau, "so unusual in a man so distinguished as was our colleague, seems then to indicate that the aptitude to speak well and to write well is not necessarily a sign of high intelligence."

At the 1889 Paris World's Fair (the one at which the Eiffel Tower made its debut), the Society of Mutual Autopsy mounted an impressive exhibit of its brains. It was wildly popular. The autopsy society, it seemed, had unwittingly created a death ritual that rivaled that of the Catholic Church. Its members had set themselves up as a closed secular priesthood, one that promised every convert the possibility of sainthood, and the chance to have their bones and brains become sacred relics. The membership forms soon flowed in, and in time, so, too, did more brains.

PAUL BROCA HAD died in 1880, at the very outset of this enterprise, and his brain was removed and placed in the collection of the Société d'Anthropologie, which happened to be where the Society of Mutual Autopsy had placed the brain of Louis Asseline. But Broca's brain was not studied. The autopsy society had no provenance over it, and it was never removed from its jar. Eventually, the autopsy society acquired its own exhibit space apart from the anthropological collection. But in 1940, with the occupation of Paris looming, the anatomical specimens from the city's museums were evacuated. After the war they were returned, en masse, to the Musée de l'Homme, which explains how Broca's brain happened to be in the company of his friends—Gambetta, Bertillon, Asseline, de Mortillet, and dozens of others—when Carl Sagan came visiting in the 1970s.

Sagan, then, had it wrong. Broca had not "established the macabre collection I had been contemplating." Nor did the collection of brains and skulls "begin in the work of Paul Broca." To be more accurate, the skulls began in the work of Gall, and the brains in the work of Rudolf Wagner.

• • •

Today the brains of the Society of Mutual Autopsy look much as they did when Sagan visited them. They are still shelved in the same row of lockers, alongside Franz Josef Gall's massive phrenological collection of plaster busts. Both collections are languishing. The brains in particular are in deplorable shape. They have not been displayed in decades, and in their current state of neglect no longer *can* be displayed. Some have broken into pieces. Others, their seals broken, have dried out completely. Fortunately, Broca's brain has been moved to the Musée Dupuytren on the left bank of the Seine, where it keeps company with the brains of Tan and Lelong.

The legacy of the Société d'Anthropologie was a mixed one. (It still exists, by the way, and still publishes its *Bulletin.*) Despite Broca's rhetorical flair, his brain weight arguments failed to carry the day, particularly outside of France, where researchers were more inclined to weigh evidence over argument. In 1880, the anatomist H. Charlton Bastian surveyed all of the available brain-weight evidence, including Broca's, and concluded that "there is no necessary or invariable relation between the degree of Intelligence of human beings and the mere size or weight of their Brains."[27] Although the idea became lodged in the popular consciousness, and was sustained by popular writers (notably Arthur Conan Doyle, in the Sherlock Holmes adventures), over the ensuing three decades it would gradually be abandoned by anatomists and physiologists.

As for localization, the credit to Broca was doled out with reservations about what he had, in fact, managed to show. It is still not clear. The faculty of speech is still not entirely understood. Brain scans have confirmed the importance of Broca's area, but cases of brain damage or surgical removal of the left hemisphere in children have also shown that a growing brain can adapt to damage to Broca's area, and reprogram other parts of the cortex to assume its role. As for the overall concept of localization, it would be advanced experimentally in the

1870s by Gustav Fritsch and Eduard Hitzig in Germany, and by David Ferrier in England, none of whom would cite Broca as an inspiration. (Perhaps this is because they focused almost exclusively on localizing motor functions by direct electrical stimulation of the cortex.)

The two most lasting legacies of the Société were the validation of localization as a paradigm of brain function (thus vindicating Gall), and the removal of the stigma of the corpse. Another legacy, a more short-lived one, was a golden age of brain collecting that lasted roughly from 1880 to 1910. Following the example of the French freethinkers, hundreds of eminent men and women, no longer squeamish at the thought of dissection, joined autopsy societies and donated their brains in the vague hope of a posthumous verification of greatness.

At the same time, there were two other types of brain donors whose brains were taken because they had no say in the matter. In order to fill out their collections, Broca and his colleagues routinely acquired brains from insane asylums and the penitentiaries. But while the French anthropologists were anxious to discover an elite brain type, they would find themselves arguing against the existence of a criminal brain. This had nothing to do with science, and everything to do with politics, and the fact that an Italian, Cesare Lombroso, had claimed the idea of the born criminal for himself.

Chapter 7

Lombroso

THE FACE OF Cesare Lombroso, the father of criminal anthropology, holds time at bay in a jar of preservative in a defunct museum of crime in the city of Turin, Italy. The museum, once a feature attraction at world expositions in the late 1800s, including the 1889 Exposition Universelle in Paris, has fallen upon hard times, and is now consigned to a few classrooms in the Pathological Institute of the University of Turin, where Lombroso once taught. It is off limits to the public, and even within this sanctum sanctorum, Lombroso's face is kept out of sight in a cluttered room on a bottom shelf behind a makeshift curtain.

It is not clear how the face manages to hold its shape. It seems to have been peeled away from the skull like a rubber mask, and has no visible means of support, yet it retains an uncanny lifelikeness, from the full goatee to the plump cheeks, and even the ears, which point outward, much as they do (or so Lombroso thought) in born criminals. The eyelids are closed, thankfully, but not the lips, as though he were a habitual mouth breather. The overall effect is that of a favorite uncle who has fallen asleep in an easy chair after a Sunday dinner.

Next to the face another museum jar, this one with a pedestal (like a large brandy snifter), features Lombroso's preserved brain. Just across from it, in an ornate glass case, Lombroso's skeleton stands at its full height, apparently just over five feet. With specimens piled next to it, in fact from floor to ceiling, there is no room in which to display anything properly. An impressive scale model of Philadelphia's Eastern State Penitentiary, for example, some eight feet across, hangs against a far wall. For Lombroso, it was the epitome of enlightened theories of incarceration.

The rest of the museum, arrayed somewhat haphazardly in two other rooms allocated to it, like a hundred Joseph Cornell boxes, conceals a good deal of method behind its seeming madness. To the knowledgeable eye it affords a tour through the life and career of a remarkable man who focused Europe's attention on the problem of crime during the late 1800s, and polarized philosophers, scientists, jurists, and politicians over his controversial theory of the criminal type. Study the criminal and his culture, said Lombroso, and you will understand the crime; punish the criminal instead of the crime, and society will reap the benefit. The idea would change basic assumptions about the treatment of convicted felons, for both good and ill, on both sides of the Atlantic for the next century. It would also set off a firestorm of controversy that swirled around the obstinate personality of a very complicated man.

"LOMBROSO IS AN ASS!" These words, spoken by an aging terrorist in Joseph Conrad's *The Secret Agent*, were directed at an audience that would have not only understood the reference, but applauded it.[1] By 1907, when *The Secret Agent* was published, Lombroso, only two years away from death, was out of favor, although his notoriety was still spreading.

"Did you ever see such an idiot?" the man continued. "And what is crime? Does he know that, this imbecile who has made his way in this world of gorged fools by looking at the ears and teeth of a lot of poor, luckless devils? Teeth and ears mark the criminal, do they?"

For Lombroso they did indeed: teeth, ears, overhanging brows, jutting jaws, and especially tattoos—all were signs of the born criminal, a creature driven by his very nature to commit horrific acts of violence. In his museum, the anthropological proof was on view—murder weapons, handwriting samples, crime-scene photographs, prisoner art and clothing, preserved patches of tattooed skin, plaster casts of faces, noses, and ears, along with piles of skulls, and, of course, stacks of brains in jars, all gathered to show that criminals make up an anthropological category unto themselves, that they inhabit a primitive culture within a modern one. Criminals are different from us, Lombroso said, and he assembled this very impressive collection to prove it.

Like Paul Broca, Cesare Lombroso is a prominent member of Stephen Jay Gould's rogues gallery, and for good reason. In the same chapter in which he exposed the fallacies of Broca's craniology, Gould took aim at Lombroso's theory of the biological basis of crime, and he scored a direct hit. Lombroso was indeed a faulty logician, a sloppy technician, and something of a mountebank. But the portrait of the man and his work is again focused too narrowly on its more outrageous aspects, which admittedly are in full view in his museum. Lombroso, it turns out, was not the same caliber of scientist as Paul Broca, although he was every bit as compassionate, and worked just as hard for social justice. Despite his mistakes, he advanced many progressive causes, including the humane treatment of the occasional criminal. Throughout a long career, he was willing to incur the indignation of colleagues, the wrath of judges, and the contempt of his fellow anthropologists (if not of novelists) in order to defend the concept that had made his reputation. He had this much in common with Franz Josef Gall: he had championed a powerful and controversial idea that seized the public's imagination, then watched it slip out of his control. Although he did not invent the criminal type, strictly speaking, Lombroso acted as a front man for the twin notions of the biological inferiority of certain races and the degeneracy of certain cultures. Ironically, as a Jew, he was vaguely aware that his ideas might be co-opted and used against his own people. Had he lived to see his theory hijacked by fascists in the 1930s, he would have been as horrified as Gould was.

He may have been wrong about many things, but like Gall, Lombroso raised useful if touchy questions that have not been fully resolved. Are criminal tendencies biologically determined? Is there such a thing as a criminal brain? The idea still has its adherents and remains controversial, but a more interesting question, one that Gould ignores even though it holds out greater hope for an answer, is this: How did such an idea catch on so quickly, and at what point did it slip out of Lombroso's grasp?

CESARE LOMBROSO WAS born in Verona in 1835 into a close-knit family that nurtured his early interest in history and philosophy. While still in public school he began writing seriously, and being taken seriously. At the age of fifteen he wrote a review of a book on the epigraphy of ancient monuments, and it so impressed the book's author that he sought out the young man and encouraged his further studies. Even as a boy, Lombroso was a dazzling writer.

Although fascinated by history, Lombroso eventually chose to study medicine, first at Padova, then Pavia, and finally Vienna, en route to earning a medical degree in 1858. He added another degree in surgery in 1859, then served for four years as an army physician in southern Italy. It was while working among soldiers in Calabria that he found his true calling—psychiatry.

Lombroso's early influences included Gall, Comte, Darwin, and the Danish scientific materialist Jacob Moleschott, a key figure in Carl Vogt's rabble-rousing circle of friends. One of Lombroso's first literary efforts was a translation of Moleschott's materialist polemic, *The Circuit of Life*. His choice of such material and his disdain for organized religion were undoubtedly fueled by the Jesuits who ran the public schools he attended as a youth. In other words, like many medical men of the era, Lombroso became caught up in positivism, anticlericalism, German materialism, evolutionary theory, and anthropological humanism at a time when each of these ideas was bursting forth on the European intellectual scene. During the 1860s, his work among soldiers, convicts, and the insane piqued an interest in psychology and psychiatry, and especially in the physiology of mental illness. Among his early works, in 1864, was a study of the relationship between genius and insanity, a topic that would preoccupy him throughout his life.

As a northerner in a country with a sharp racial divide, Lombroso used southern Italy, particularly Calabria, as his laboratory. While there, he studied cretinism (a developmental disorder) and pellagra (an inflammatory disease) among the local population. He also became fascinated, if not obsessed, with the phenomenon of tattooing among soldiers and convicted felons. Lombroso associated tattoos with primitive cultures, and he was shocked by the pornographic imagery favored by hardened criminals. "From the very beginning," he wrote, "I was struck by a characteristic that distinguished the honest soldier from his vicious comrade: the extent to which the latter was tattooed and the indecency of the designs that covered his body."[2]

In the early 1860s, Lombroso took part in a campaign by the Italian army to round up roving bandits who had been plaguing the south during the period of Italian unification. It was on this campaign that Lombroso met the most famous of the brigand leaders, Giuseppe Vilella, who fit the classic profile of the charismatic warlord—witty, cynical, and overflowing with boastful arrogance. The encounter proved to be a turning point in Lombroso's life. After Vilella's death in 1871, he managed to acquire the brigand's skull, and, while contemplating it, he was struck by a revelation. "At the sight of that skull, I seemed to see all of a sudden, lighted up as a vast plain under a flaming sky, the problem of the nature of the criminal—an atavistic being who reproduces in his person the ferocious instincts of primitive humanity and the inferior animals."[3]

The feature that caught Lombroso's eye was a depression on the inside of Vilella's occipital bone (at the very back of the skull). He could not recall having seen this feature in any human skull, although it was fairly typical in animals, especially in "the lower types of apes, rodents, and birds." He named it the median occipital fossa, and he decided that it was the sign, if not the source, of Vilella's violent tendencies. Thus was born Lombroso's theory of the biologically determined criminal.

The skull of Vilella became the first and most important specimen in Lombroso's museum of criminal anthropology, and the basis for his anatomical theory of crime. If it all sounds a bit too neat, that's because it is. It turns out that Lombroso embellished if not invented the Vilella episode years after the fact in order to provide himself with a suitable creation myth. As with Broca's presentation of the brain of Tan, or Gall's reminiscence of his school chum with the bulging eyes, the story serves a literary purpose that hides a more complicated truth: Lombroso was not so much an innovator as a popularizer. The theory of the born criminal had been floating around for some time before he decided to make it his own.

The list of Lombroso's ideological forebears is too long to include here. Gall is an obvious example. (His list of mental faculties included "murderousness," which Spurzheim later changed to "destructiveness.") The physiognomist Johann Kaspar Lavater, who read the signs of aggression in the shapes of faces, is another. One of Lombroso's favorite markers of criminality, the sloping forehead, dated back to the late 1700s, when Peter Camper, a Dutch anatomist, introduced the Camper angle, a measure of deflection of the skull's frontal bone. Even young Eugène Dally of the Société d'Anthropologie had examined the hereditary and organic nature of insanity and criminality before Lombroso got into the game.

According to the social historian Marvin Wolfgang, Lombroso "fell heir to the growing body of medical, clinical, and psychiatric literature that dealt directly and peripherally with the criminal. *From* this knowledge he gained a perspective and theoretical orientation: *to* this knowledge he added new, exciting and controversial dimensions."[4] What he added was a catalogue of atavistic features—twenty-two of them in men and twenty in women—that marked the born criminal as surely as a broad vertical forehead marked a man of great intellect. In essence, Lombroso combined Lavater's science of physiognomy with Gall's phrenology to create a diagnostic method for identifying society's born losers.

IN 1844, KARL Marx and Friedrich Engels, then living in Paris, remarked with some alarm on the ease with which violent criminals could wreak their havoc and then disappear into the winding alleyways of the city. During the first half of the nineteenth century, Paris's population had doubled, as a large transient group of unemployed peasants overran districts that had been laid out centuries earlier on a labyrinthine plan. In short order these neighborhoods devolved into Hobbesian rat's nests, where life was nasty, brutish, and short. Novelists like Eugène Sue, Victor Hugo, and, later, Émile Zola, may have romanticized this criminal underworld, but city dwellers viewed it as a plague. The criminals of Paris, it seemed, had found a protective environment, one that nourished and sheltered them; the only solution, it seemed at first, was an enlightened urban design that eliminated the protection afforded by darkness. Marx and Engels were pleased to note that "at this very moment broad light streets are being laid out in the Cité to give police access to them."[5] But such measures addressed the symptoms rather than the causes of the problem. Something more had to be done.

The first researcher to take a comprehensive view of the problem of urban crime was Adolphe Quetelet, a Belgian mathematician and sociologist who had moved to Paris in the 1820s in order to study astronomy, and possibly to become a playwright. Quetelet soon fell under the spell of Pierre-Simon Laplace, the mathematician and contemporary of Carl Gauss who had taken Gauss's bell curve beyond physics into the realm of human attributes and behaviors. While Laplace was content to dabble in the broader implications of the law of errors, Quetelet became a data-collecting dynamo, and single-handedly invented a field of study that he called "social physics"—an application of statistics to human populations that now goes by the name of demographics. His first topic of study was crime.

Quetelet looked at murder rates in Paris over a period of years and broke them down by category. He found that not only did the number of murders from year to year remain fairly constant, but so, too, did the number committed by knife, by sword, by strangulation, by poison, and so on. Quetelet concluded that although a single murder could not be predicted, murder as a phenomenon seemed to follow an unvarying mathematical pattern, and could be studied by scientific methods. Each year, the number of killings by knife, for example, was a function of the number of people crammed into the city and the conditions under which they were forced to live. Alleviate the causes, he wrote, and the number of stabbings would drop.

Quetelet's most influential work, *On Man and the Development of His Faculties,* appeared in 1835. It contained two famous ideas. The first was his belief that criminals are products of their environment, that crime is essentially a social problem. In the absence of social remedies, he argued, crime statistics would remain constant from year to year. The second proved to be Quetelet's most lasting invention. He called it *l'homme moyen*—the average man—a statistical construct that remains popular to this day.

Every measurable physical and intellectual human attribute, Quetelet found, conforms to a bell-shaped distribution. The ideal, taking the aggregate of all possible bell-shaped curves illustrating every imaginable variable, occurs at the mean, or at the peak of the bell. At the confluence of all such means stands the average man. We are all, he said, deviations from the "normal" in various ways, and in that sense the average man is a pure fiction. But for Quetelet, the mean (or average) also represented perfection.

In the 1870s, Italy presented a very different set of problems than those Quetelet studied in Paris. Having just achieved independence from Austria, having barely made the transition to nationhood, Italians faced a crime problem that had less to do with overcrowding than with sheer confusion. Their new country was little more than a concept. In its first thirty years of existence, it would go through twenty-eight governments. Its previously autonomous provinces, each possessing its own dialect, constituted a nation divided (as is still said of Great Britain) by a common language. At the time of unification, less than 1 percent of its inhabitants spoke the national language; up to three-quarters were illiterate. There was no such thing as an "average Italian." Southern Italy in particular was plagued by brigandage and a growing anarchist movement (fomented in part by Carl Vogt's friend Mikhail Bakunin, who had taken up residence in Naples). It was one thing to make a unified Italy, but quite another to make Italians out of its residents. As Lombroso himself admitted, "Even in evil, Italy is not bound together."[6]

It is not surprising, then, that Lombroso would take a different view of crime than Quetelet, or that he would focus on the average criminal rather than the average man. As he saw it, it was not crimes that fell into predictable categories, but the men and women who committed them.

AT THE OUTSET of his career, Lombroso had ample opportunity to study criminals. He carried out his first anthropometric studies by using Broca's methods on the soldiers with whom he served in the early 1860s. He then applied the same methods to the criminally insane while serving as director at a succession of asylums over the next decade. By the mid-1870s, he had also gained access to several prisons, where he collected the data that resulted in such early papers as "Anthropometry of 400 Venetian Criminals" and "Emotions and Passions of Criminals." At the same time Lombroso began to delve into the culture of prison life. He saw himself, in the tradition of Broca, as an anthropologist collecting data from the field, assembling a museum of folkloric artifacts, as well as the evidence that would eventually add up to a theory. The result, in 1876, was one of the most influential books ever written on crime—*L'Uomo delinquente (Criminal Man).*

According to Lombroso, the typical criminal is an evolutionary throwback, a savage, if not an animal. The evidence is written into his anatomy. "Thus were explained anatomically the enormous jaws, high cheek bones, prominent superciliary arches, solitary lines in the palms, extreme size of the orbits, handle-shaped ears found in criminals, savages, and apes, insensitivity to pain, extremely acute sight, tattooing, excessive idleness, love of orgies, and the irresponsible craving of evil for its own sake, the desire not only to extinguish life in the victim, but to mutilate the corpse, tear its flesh and drink its blood."[7] (The description, except for the tattoos, is essentially that of Count Dracula, whom Bram Stoker modeled directly after Lombroso's criminal type.)

To make his theory stick, Lombroso had to make the case that the animal world, which has a greater claim on criminals than the civilized world, is by its very nature violent, amoral, and predatory. Thus he begins his book with a catalogue of criminal behavior in animals and plants. Although he considered himself a positivist scientist, Lombroso's method was anything but scientific. His signature mode of argument was to seize upon an idea such as the ferocity of animals or the confluence of genius and insanity, then find as many instances of it as he could in the literature, and list them in paragraph after paragraph of authoritative rambling. To read Lombroso is to get the impression that there are no counter examples, or that if there are, they exist merely to prove his point, because as everyone knows, there is no rule that does not admit exceptions.

Criminal Man went through five editions over the next twenty years, with each new edition incorporating Lombroso's revisions in the face of criticisms not only from opponents (most of them French), but from his own followers. From a first edition of 252 pages in 1876, the book would grow to 1,903 pages in two volumes in the final edition of 1896. Although it did not solve Italy's crime problem, *Criminal Man* succeeded in making Lombroso a household name, one of the most talked-about public intellectuals in Europe.

IT IS EASY to get caught up in the idea of Cesare Lombroso as a fascist: a scourge of the poor, the illiterate, the downtrodden, and the racially impure. But this would be a mistake. Lombroso is not innocent of Gould's charges of faulty reasoning, but before condemning him outright, it is useful to consider the circumstances that gave rise to Lombrosoism and allowed it to flourish as an idea (because it did not flourish as a social program). After all, Lombroso invented the idea that punishment should be meted out with due consideration of motive and intent, and he deserves the same consideration.

Lombroso's theory of crime and punishment filled a century-old void in Italian legal philosophy. It presented the first alternative to the so-called classical school, founded in 1764 by the Milanese economist Cesare Beccaria. Beccaria's book—*On Crimes and Punishments*—is still considered to be the most important criminal justice treatise ever written. It went through eight editions within just a few years, was translated into every European language, appeared in an American edition in 1776, and was incorporated into the penal codes of most Western nations.

Beccaria's slim volume rescued republican governments from the prevailing medieval approach to crime, and it established its author as the most important legal commentator of the Enlightenment. Before Beccaria, the penalties for serious crimes were meted out at the whim of magistrates and rulers. Trials were often held in secret. The range of punishments included mutilation, public humiliation, torture, forced labor, and, of course, brutal executions. "In order that any punishment should not be an act of violence committed by one person or many against a private citizen," Beccaria wrote, "it is essential that it should be public, prompt, necessary, the minimum possible under the given circumstances, proportionate to the crimes, and established by law."[8]

Beccaria's theory, grounded in the paired assumptions of free will and individual responsibility, held all citizens equally responsible before the law, and equally accountable for the consequences of their actions. Intention, he said, should have nothing to do with the severity of penalties. Punishments should fit the nature of the crime, Beccaria argued, and they should take no account of the circumstances, the status, or even the nature of the perpetrator. The guiding principle of Beccaria's system was deterrence above all. It seized upon the utilitarian notion of the greatest good of the greatest number, and in doing so focused on the crime rather than the criminal.

Lombroso took a longer view of the problem, and he decided that Beccaria had it backward. Instead of deterrence, dangerousness should be the focus of law enforcement. Not merely the intent but the very nature of the criminal had to be taken into account. Some criminals—occasional criminals—did not represent a great threat to their neighbors. They were unlikely to commit another crime, and their punishment should be light. The habitual criminal, on the other hand, would not respond to any form of deterrence or reform. The only effective way to protect the populace from such types was to remove them from society altogether.

Lombroso's most enduring legacy was his recommendation of open-ended (or indeterminate) sentencing, a policy that would eventually catch on in many Western countries, particularly in the United States. It proposed that courts should deal with criminals in the same way that physicians deal with patients, on a case-by-case basis, taking the entire medical (or criminal) history into account. If crime was indeed a disease, then the criminal was, in essence, a microbe, and some microbes were more infectious and deadlier than others. What Lombroso offered was a way for the average citizen to protect himself by recognizing the difference, to pick out the unsavory types at a glance. The police, if properly trained, could do even better. The potential abuses of such a subjective method of diagnosis are obvious enough, but would not be enacted into law until the 1930s, when Benito Mussolini promulgated a series of racial laws based partly on Lombroso's theory of biological inferiority, and began to round up the Jews.

Lombroso is hardly blameless for the fallout from his theories, although, to be fair, many of his ideas were enlightened and beneficial. Politically, Lombroso was a liberal, a socialist, and a champion of the poor. He preached an end to feudalism and argued for the redistribution of the land among the peasantry as a way to keep them from resorting to crime. To the same end, he campaigned for improvements in education, public health, and hospitals. He also explored every possible alternative to incarceration, which he recommended only for serious or incorrigible offenders. He studied innovative prison designs, particularly the cell-block penitentiary (like the one in Philadelphia), which isolated prisoners and minimized the spread of prison culture. Early in his career he opposed the death penalty; he changed his mind only when confronted with perpetrators of the most brutal acts.

If Lombroso had a tragic flaw, it was (naturally) hubris. He was obsessed with the idea of the criminal type, with the conceit that he had stumbled upon something truly revolutionary. For this reason it is difficult to assign Lombroso to any school of thought but his own. He had read Darwin, but he was not, strictly speaking, a Darwinist. Like Broca, after reading *On the Origin of Species* he dropped his belief in polygenism—the separate creation of the races—but he retained the biological basis of racial classification, and a hierarchy of racial types. He called himself a positivist, but he was not a follower of Comte. He subscribed to positivism in the vague sense of the word that most scientists did at that time: he believed in the sanctity of data and the redemptive power of empirical proof. Although influenced directly or indirectly by many thinkers—among them Gall, Comte, Quetelet, Broca, and Darwin—he swore allegiance to none of them. He was his own guru, and his downfall would be his refusal to share the stage with anyone.

THE THEORY OF the criminal man was appealing for nonscientific reasons. Like most successful pseudoscientific paradigms, it was easy to grasp, it provided a simple explanation for a troubling facet of life, and it had one other overriding strength—the force of its creator's personality. Lombroso was an engaging writer, a relentless popularizer of his big idea, and an indefatigable collector of evidence—scientific, historical, and even anecdotal—whose sheer scope and range were matched at times only by their absurdity and irrelevance. He was not above enlisting fictional characters as case studies of the born criminal. Thus Shakespeare's Macbeth and Dostoyevsky's Raskolnikov became exemplars of epileptic depravity. Unmethodical, capricious, and slipshod are just a few of the charges that were leveled against him, yet no one could deny that he was disarmingly earnest and entertaining. (Even so, when he visited Leo Tolstoy in 1897, the cantankerous novelist was unimpressed. "Lombroso was here," he wrote, "a limited naive oldster.")[9]

As a scientist, however, Lombroso did not get a free pass. His worst habit, if not his signature rhetorical device, was his refusal to acknowledge the possibility that common underlying factors could explain the relationship between two seemingly connected phenomena. In other words, he routinely confused correlation with causation. Thus, for Lombroso, atavism caused criminal behavior. It was neither coincidental nor (as in the case of tattooing) easily explainable by another factor (that tattooing, for example, was merely a part of prison culture). Like Gall, Gratiolet, and even Broca, Lombroso tended to extrapolate general rules from isolated cases, and he read too much into extraordinary pieces of evidence, such as the skull of Giuseppe Vilella. Like Carl Vogt, he insisted that exceptions only proved the rule. Although he understood the importance of control groups, he rarely tried to assemble them. But perhaps most damning of all, he never defined the very terms he proposed to demystify. He even failed to produce working definitions of the crucial terms *crime* and *criminal.* For Lombroso, criminals were simply the inhabitants of prisons, and crimes were the reason they had come to be there.

Most of this is not surprising. Lombroso was not an anatomist by training, nor an anthropologist. He was a physician, had served in the campaign against brigandage in the early 1860s, and then taken charge of an insane asylum in Pavia. Like all psychiatrists of the era, he learned his craft on the job. His lack of statistical training is evident in all of his works, which could serve as textbook examples of the worst abuses of the statistical method.

When it came to the study of brains and skulls, he was completely out of his depth. He collected them by the hundreds—they were readily available. He argued that among the atavistic characteristics of criminals were thick skulls (of the kind possessed, as it turned out, by Gall and the anatomist Friedrich Tiedemann) and underdeveloped brains. His discovery of the occipital depression of Vilella's skull was regarded so lightly that few people bothered to challenge it. Lombroso did perform hundreds of autopsies, and he and his students preserved thousands of anatomical specimens, but he was, like many elite brain researchers, something of an amateur, as well as, apparently, a body snatcher.

Lombrosoism might even have quietly faded away had it not been for Broca's Société d'Anthropologie, which made the mistake of taking Lombroso too seriously. When Broca died in 1880, his protégé, Paul Topinard, assumed control of the Société. Along with the anthropologist Léonce Manouvrier, Topinard took up the issue of the born criminal in the wake of Lombroso's debut. They were joined by the criminologist Alexandre Lacassagne and the sociologist Gabriel Tarde (incorrectly singled out by Gould as a colleague of Lombroso's), in what became known as the French school of criminology. Together, they launched a full frontal assault on *Criminal Man* and its insistence on the biological, as opposed to environmental, causes of most crimes. The French, despite being militant materialists, were not willing to accept the biological basis of *all* behavior. Preferring to view society itself as an organism, they turned to Auguste Comte's sociology for answers.

It was in response to this that a so-called Italian school of criminology closed ranks around Lombroso. Prior to the French attack, there had been little enthusiasm for Lombroso in Italy. But now even his critics rallied to his defense, and the battle lines were set.

The first skirmishes in this little war played out at a series of international congresses of criminal anthropology, the first of which met in Rome in 1885. Lombroso dominated the proceedings and harped on his theory of atavism to the point of ignoring environmental causes of crime altogether. As if to drive home this point, the Italians mounted an elaborate exhibit featuring anatomical specimens donated by over forty participants. (Lombroso contributed hundreds of items, including tattooed skin, casts of ears, and, of course, Vilella's skull.) These gruesome exhibits made such an impression that the French contingent immediately found themselves on the defensive.

At the second congress, which met in Paris in 1889, the counter-attack was led by Tarde, who tried to demolish Lombroso's claims to scientific objectivity by demanding to know exactly how many atavistic stigmata were necessary for a diagnosis of born criminality. Who is a criminal? he asked the Italians. Where were the control groups? How can an objective conclusion be drawn from variables—physical traits—that are judged so subjectively?

Topinard and Manouvrier were also skeptical of Lombroso's data. They had done their own studies of criminal brains (as had Broca), and instead of finding them to be smaller and less developed than noncriminal brains, as Lombroso alleged, they found them to be larger. Broca had attributed this result to the advantages of youth—most criminals die young, before their brains have had a chance to atrophy. Death by hanging was also thought to engorge the brain with blood, producing deceptively high brain weights. In Munich, the anatomist Theodor Bischoff, who had collected over a hundred brains of executed criminals, published a study in which he reached the same conclusion.[10] Not only were these brains heavier than those of law-abiding citizens, but a few criminals had brains of Byronic proportions. And yet such was the state of brain research that Lombroso could point to at least one study that came to the opposite conclusion.

In 1878, Moritz Benedikt, a Hungarian anatomist living in Vienna, published *Anatomical Studies on the Brains of Criminals,* a work based on twenty-three case histories of men who were executed or died in prison. Coming hot on the heels of *Criminal Man,* it provided Lombroso with seemingly irrefutable evidence of the existence of a criminal brain. In almost all instances, Benedikt noted a unique feature in the surface configurations of his specimens. Each one showed a pattern of "confluent fissures," which Benedikt took to indicate a deficiency in the development of the gyri (the ridges) leading to an excess of sulci (the fissures). In layman's terms, in the cerebral hemispheres of criminal brains the fissures all ran together like a series of interconnected canals. "If we imagine the fissures to be water-courses," wrote Benedikt, "it might be said that a body floating in any one of them could enter almost all the others."[11] Benedikt concluded that because of their inferior brain morphology, "criminals are to be viewed as an anthropological variety of their species, at least among the cultured races." He claimed that this discovery would create a revolution in ethics, psychology, jurisprudence, and criminalistics.

It might have, had Benedikt possessed a fraction of Lombroso's literary flair. Instead his study makes for dull reading. In paragraph after paragraph, he describes fissures that extend too far or not far enough, ridges that are too robust or too insubstantial. If the descriptions of the brains are repetitive (which was, after all, Benedikt's point), the stories of the men themselves, their crimes, their fates in prison, the "complete shipwreck of their lives," are not. Twenty-five-year-old Vaso Zatezalo, for example, died while serving a five-year term for manslaughter, having "ripped open his antagonist's abdomen." "The man was addicted to drink," notes Benedikt, "had weak mental powers, and possessed the lowest grade of cultivation." As for the brain: "Gyrus lingualis thick and short; and entire occipital basilar lobe cut up into islands."

Benedikt's book would probably have been forgotten had it not been translated into English by the American phrenologist E. P. Fowler in 1880, and disseminated among a generation of physicians who were intrigued by the possibilities of phrenology. In Montreal, a young pathologist named William Osler reviewed Benedikt's book and decided to conduct a study of his own by comparing the brains of executed convicts with those of patients who had died at Montreal General Hospital.[12] Although he could neither confirm nor refute Benedikt's theory, he thought it plausible enough to merit further investigation. From a philosophical standpoint, however, he was deeply troubled. So were the French.

The French school of anthropology had emerged out of an unsentimental combination of materialism and atheism that denied the existence of free will. This left them in a difficult position. How could they assign blame to any criminal without giving in to Lombrosoism?

What degree of responsibility should or could be attributed to someone with a defective brain? Lombroso assigned none, and advocated the removal of the habitual criminal from society. As a literary man himself, Osler turned to Shakespeare and quoted Iago to the effect that " 'Tis in ourselves that we are thus and thus. Our bodies are gardens to which our wills are gardeners."[13] Osler concluded that it would be a mistake for society to accommodate a class of "criminal automata." Instead, the law should continue to be a "terror to evil-doers." In other words, Osler sided with Beccaria and argued for the value of deterrence. Like the French anthropologists, he favored social remedies for crime.

Among those who rallied around Lombroso at the 1889 Paris congress were his former students, Enrico Ferri (the man who had coined the term *born criminal* in 1883) and Raffaele Garofalo. They were joined by Moritz Benedikt and the materialist agitator Jacob Moleschott. In response to stinging criticism from Tarde, Lombroso's team issued a challenge. They asked that a commission be appointed to carry out a comparative study that would tally atavistic characteristics in a hundred criminals and a control group of the same size. Among the seven men appointed to the commission were Lombroso and Benedikt from the Italian school, and Lacassagne, Manouvrier, and Alphonse Bertillon (the son of the Society of Mutual Autopsy's Louis–Adolphe Bertillon) from the French school. The results, to be presented at the third congress in Brussels, would settle the matter. Unfortunately, the idea was dropped, the Italians boycotted the 1892 meeting in protest, and the question was never resolved.

Despite their unified opposition to the French, the Italian criminologists remained critical of Lombroso. His leading disciple, Enrico Ferri, criticized him for attributing far too high a percentage of crime—up to 70 percent—to born criminals. At Ferri's urging, Lombroso gradually reduced this proportion in successive editions of his book to a mere third. He did this partly by expanding his classification of criminal types so that it eventually came to include the epileptic criminal, the insane criminal, the occasional criminal, and the habitual criminal. In expanding his list of classifications, Lombroso seems to have taken a page out of the phrenologists' playbook. Instead of admitting he had been wrong, he cleverly amended his theory in order to head off his critics.

Like Gall, Lombroso had a maddening tendency to argue from conclusion to premises. Here is a criminal, he would say: Observe the atavistic features in his face, in his skull, the shape of his ears, in the frontal convolutions of his brain. (Lombroso, of course, provided the inspiration for Dr. Waldman's speech at the beginning of *Frankenstein.*) But to be of any scientific use, the procedure had to be invertible. Could Lombroso look at a lineup of men and say which one was a criminal, or scan a row of brains and say which one was a criminal's brain? He could not. But with a little tweaking of his theory, he found that he could easily sidestep the question altogether.

IN THE 1850S, a young physician named Bénédict Augustin Morel arrived in Paris and secured a position at the Salpetrière, France's largest insane asylum, where he became an alienist. An alienist is someone who studies the alienation of the mind from a state of composure. The current term of the art is *psychiatrist,* and Morel would distinguish himself among psychiatrists by identifying one of the most severe forms of alienation, and coining a term to describe it—*dementia praecox.* In 1908, the affliction was given the name it goes by today: schizophrenia.

Morel's other legacy has managed to survive in the language partly through the efforts of Cesare Lombroso. It even made a comeback in pulp fiction of the 1950s. This was his discovery of the *degenerate,* a term that would eventually be applied to depraved criminals and juvenile delinquents, although for Morel, degeneration went deeper than teenage angst. It was, he claimed, a hereditary condition that was closely linked with vice.

Morel's theory of degeneration began with his observations of cretins—shepherds and other residents of mountainous regions who suffered from a peculiar form of mental and physical retardation. Morel attributed the condition to a compromised line of descent whose source could be traced to a variety of vices, notably alcohol, tobacco, and opium, or to high-risk behaviors that resulted in venereal disease, tuberculosis, and goiter. Morel argued that insidious habits damage not just the individual, but also his or her descendants, and that the effects were cumulative. After several generations they resulted in extreme mutations such as cretinism, or worse. What Morel did not know was that cretinism was caused by an iodine deficiency, essentially by a lack of fish in the diet.

Morel did not invent degeneration. J. F. Blumenbach, Rudolf Wagner's predecessor at Göttingen, had used the term in a nonjudgmental way to describe how racial characteristics arose in response to environment. In Blumenbach's usage, to "degenerate" simply meant to diverge from the original stock, and had nothing to do with deterioration. The racist connotation emerged when monogenists used degeneration to explain the inferiority of the African race as a failure to attain the same degree of improvement as the white race. (This was one of the aspects of monogenism that would turn Paul Broca, at least initially, away from a belief in a common origin of the races.) Morel's concept of degeneration, by contrast, did not address race at all, but focused instead on intelligence and moral sense. It was, in essence, an explanation of society's ills. The worst of these, of course, was crime.

Morel's *Treatise on the Physical, Intellectual, and Moral Degeneration in Humans* appeared in 1857, two years before Darwin's *On the Origin of Species.* While book and author would soon recede into oblivion, the idea of degeneration would take on a life of its own. Like positivism, it provided a useful catchall. In particular, it allowed Cesare Lombroso to expand his classification of the criminal type.

Degeneration is, in essence, the encroachment of biological determinism into the realm of free will, the subjugation of choice to the dictates of animal nature. The idea relies on a mechanism of genetic transmission that was widely accepted in Lombroso's time (even Darwin subscribed to it), although it was later to be proved false by Gregor Mendel. This was Lamarckism, named after Jean-Baptiste Lamarck, a zoologist and colleague of Georges Cuvier.

Lamarckism is an evolutionary theory that proposes the inheritance of acquired traits. It is most compellingly illustrated by the kinds of just-so stories favored by Lamarck himself. How the giraffe got its long neck, for example, is the story of hundreds of generations of giraffes, each of whom reached just a little bit higher and a little bit farther than the next giraffe, and so caused a barely perceptible extension of its neck. This trait, once acquired, was then passed on to the next generation, and bit by bit, over thousands of years, the modern giraffe evolved. (In Darwinian evolution, chance variations produce some giraffes with longer necks than others. Longer-necked giraffes, being better suited to compete for food, enjoy a higher survival rate. They do not elongate their necks, but instead have more success at passing along their long-necked genes.)

In other contexts, particularly in the matter of intelligence, perhaps the most compelling of acquired characteristics, the argument can be far more subtle and persuasive, to the extent that Lamarckism still has its adherents to this day. (The white race, in the just-so story advanced by Lamarckian scientific racists, expanded its intelligence generation by generation, in the same way that generations of giraffes had extended their necks, until whites loomed over the other races.)

Morel used Lamarckian evolution to argue that physical stigmata and physical infirmities invaded the body so profoundly as to alter the mechanism of genetic transmission. All acquired diseases and deformities became heritable traits. The only good news for the human race, as far as he could see, was that these acquired weaknesses would gradually be selected out of the population because they were accompanied by infertility and high mortality. While they lived, Morel suggested, the most humane thing to do with victims of degeneration was to treat them as well as possible in mountain spas and retreats, and wait for their kind to become extinct. Lombroso had the same idea, although he recommended prisons instead of spas.

In the third edition of *Criminal Man,* Lombroso included degeneration as a cause of criminality. It would prove to be his trump card. Degeneration enabled him to connect criminal behavior with visible atavistic traits. In the original version of his theory, Lombroso viewed the born criminal as an evolutionary throwback, a case of arrested development on the evolutionary scale. But it was Morel's degeneration theory that provided him with a justification for the genetic basis of crime. To explain the connection, Lombroso had one other tool at his disposal, yet another prevalent belief of the time (this one also subscribed to by Darwin, and even Freud). It was called recapitulation.

Originated by the German embryologist Ernst Haeckel in 1866, recapitulation states that "ontogenesis is a brief and rapid recapitulation of phylogenesis." Loosely translated, this means that in the passage from embryo to adulthood (ontogenesis), every organism replays the evolutionary stages of the development of its species (phylogenesis).[14] According to recapitulation theory, the tadpole is essentially a prehistoric frog. The human fetus, at some stage, exhibits several fish-like features, betraying man's prehistoric ancestry in ocean dwellers. The theory also implies that children behave, on a moral level, like savages because they literally are savages—humans at an early stage of evolution. It seemed natural to assume that some children never fully matured, and thus became stranded on a lower rung of the evolutionary ladder. These were Lombroso's atavistic throwbacks.

As a paradigm, recapitulation provided a useful explanation for abnormal or asocial behavior, and, in particular, for crime. But not all crime. In order to respond to his critics, Lombroso had to come up with a biological basis for criminal behaviors other than those attributable to atavism. Degeneration would provide the answer.

In his 1857 treatise, Bénédict Morel indulged in the classification of degenerative types. The epileptic was one type; the insane or mentally deficient was another. Criminals were also degenerates. It was Morel, in suggesting that the study of degeneration fell within the realm of "morbid anthropology," who implied that many degenerate traits were visible only to the pathologist, who could look for the evidence in lesions and other internal anomalies during an autopsy. This provided Lombroso with the capstone of his grand theory. For those criminals who did not show any outward signs of degeneration, he could point to the physiology of their brains or even to individual cells. Some born criminals, he said, showed no outward stigmata at all. They committed crimes because they suffered from a form of biologically based moral degeneration that was invisible to the untrained eye.

In the fourth edition of *Criminal Man*, published in 1889, Lombroso added epilepsy to his list of criminal traits. Although it was not more common among criminals than the general population, Lombroso came to regard it, whether in its full-blown form characterized by spastic fits, or in a latent and less-detectable form, as a pervasive degenerative condition. This allowed him to reapportion the amount of crime attributable solely to the born criminal. "The fusion of criminality with epilepsy and moral insanity alone," he wrote, "could explain the purely pathological and non-atavistic phenomena in the delinquent."[15] Whether they manifested the symptoms or not, all born criminals, as far as Lombroso could tell, were epileptics. Eventually he would come to the same conclusion about another degenerate type, one that had fascinated him from the very outset of his career. This was the born genius.

"IT IS A sad mission to cut through and destroy with the scissors of analysis the delicate and iridescent veils with which our proud mediocrity clothes itself."

The opening sentence of his 1891 book, *The Man of Genius,* gives a hint of Lombroso's infectious literary style, infused by the sensibility of a pathologist. The sentiment is decidedly materialist. "Very terrible is the religion of truth," Lombroso regrets to inform. "The physiologist is not afraid to reduce love to a play of stamens and pistils, and thought to molecular movement. Even genius, the one human power before which we may bow the knee without shame, has been classed by not a few alienists as in the confines of criminality, one of the teratologic forms of the human mind, a variety of insanity." The leading voice among these alienists, of course, was Lombroso himself.

The Man of Genius consists of four hundred pages of anecdotes, folklore, charts, graphs, and capsule biographies that drive home the idea that "people of genius are not only unpleasant in practical life, but weak in moral sense and wicked." The quote is from the philosopher Arthur Schopenhauer, who merits several pages in Lombroso's book as the best example of "the most complete type of madness in genius" there ever was.[16]

In a chapter devoted to the skulls and brains of men of genius, Lombroso catalogues cephalic abnormalities in men such as Pericles, Dante, Machiavelli, Kant, and Alessandro Volta. "In Volta's skull," he writes, "I have noted several characters which anthropologists consider to belong to the lower races." These included an "obtuse facial angle" of seventy-three degrees, the infamous Camper angle.[17] A receding forehead, one of Lombroso's atavistic traits, was a sure sign of a deficient frontal lobe, the supposed seat of the intellect. A vertical forehead, on the other hand, was the classical ideal, and yet Volta had managed to overcome this deviation from the classical, and made discoveries worthy of genius. How had he done it?

Lombroso was not bothered by inconsistencies. Such exceptions were perfectly consistent with his observation that geniuses may excel at one thing and come up woefully short at others. Volta, in Lombroso's view, must have been a freak of nature.

As in so much of Lombroso's writing, a recitation of famous cases accompanied by page after page of lists and diagrams creates a vague impression of objectivity, while obscuring the fact that the author never states a thesis. What he does show is that it is easy to find descriptions of famous brains and skulls that deviate from the "normal" in one way or another. What he fails to admit is that *all* brains and skulls deviate from the normal. Or, more to the point, that there is no such thing as a normal brain.

It should be noted that Lombroso did not collect brains of eminent men. He had very little access to them. Other than Giuseppe Vilella, his only brain-anatomical diagnosis of genius occurred in 1903, after the death of his colleague and nemesis Carlo Giacomini. Unlike Lombroso, Giacomini was an important brain anatomist with an international reputation. Throughout a prolific career at the University of Turin, he often found himself obliged to refute Lombroso's claims about the existence of criminal brains or of so-called "race brains." Giacomini showed, for example, that there is far more variation in brain morphology within races than between them, a fact that makes it impossible, a priori, to distinguish the brain of a black man (or woman, for that matter) from that of a white man, or to diagnose atavism in a brain. When Giacomini died, his students performed, at his own request, a postmortem examination in which they preserved his skeleton, brain, viscera, and also his face. (These are currently on display at the Institute of Human Anatomy in Turin, and may have inspired Lombroso to leave similar instructions for the disposition of his own corpse.) At the presentation of Giacomini's autopsy results, Lombroso created a scene when he stood up in the middle of the crowded lecture hall, raised his finger in the air, and announced that he had found the median occipital fossa in Giacomini's skull, a clear sign of epileptoid degeneracy, or, in this case, the mark of a genius. (Although the feature seems to appear randomly in about 4 percent of the population, Lombroso's own skull, it was later revealed, had the very same occipital depression.)[18]

For Lombroso, the connection between genius and insanity went much deeper than brain structure. He sought its causes in climate, barometric pressure, and geography. According to his statistics, rising temperature and air pressure cause an increase in manic attacks; hilly regions are conducive to the composition of poetry; and May is the most propitious month for artistic creativity.

In a final chapter, Lombroso took five cases of supposedly sane men of genius—Galileo, Leonardo da Vinci, Voltaire, Michelangelo, and Darwin—and showed that degenerative psychoses often go unreported. Darwin, he noted, was surely a "neuropath" who spent many days incapacitated by vague illnesses; Michelangelo was "affected by a neuropathic condition bordering on hysteria."[19] For Lombroso, the brain of a true genius, far from being superior to a "normal" one, was almost always defective.

FOR ALL OF the hand-wringing over Lombroso's legacy, he would have to be counted a failure as a social reformer. Very few of his suggestions were enacted into law. His greatest defeat occurred with Italy's adoption in 1889 of the Zandarelli penal code, which was Beccarian in spirit. It banned the death penalty (which Lombroso, in an odd betrayal of his liberalism, had come to support). It did make concessions to the circumstances of a given crime, while maintaining the priority of the crime over the criminal. Perhaps Lombroso's greatest legacy is in the area of sentencing, particularly in the United States, where his recommendations for indeterminate sentences, for early release, house arrest, suspended sentences, fines, and, above all else, for taking the intent of the perpetrator into account, were all incorporated into the penal code, although not until well after his death.

As for his attempt to have anthropometric data entered into evidence at criminal trials, Gould cites one instance in which Lombroso appeared as an expert witness at a murder trial and argued for conviction in the absence of any physical evidence beyond the accused's appearance. The man was acquitted. This type of expert testimony did occur in Italy, but it never gained a foothold, given that judges did not relish the prospect of relinquishing their authority to a bunch of anthropologists.

Lombroso died in 1909 and, like Giacomini, was dissected and preserved by his students, who deposited his skeleton, brain, and face in his own museum. Had nothing else come of his work, he might have quietly slipped into obscurity. But he had planted a seed, and thirty years later his grand theory degenerated into an evil parody of itself when Benito Mussolini enacted a series of racial laws based in part on Lombroso's theory of atavism. Mussolini effectively solved his Southern Question by declaring that all Italians belonged to a pure race that had not been contaminated by interbreeding in over a thousand years. He left Gypsies and Jews out of this equation so that they could take the place of southern Italians as the degenerate population of the moment. They were forbidden to intermarry, to teach, to work for the state, or to own property. Eventually they were deported to death camps.

Whether this would have happened without the legacy of Lombroso is difficult to say. Most scholars agree that he helped to set the stage for the eugenics movement (even though Lombroso himself never advocated eugenic measures), from which it was a short step to genocide. Yet it is not as if Gould, in 1978, was the first writer to debunk Lombroso's theory and methods. During his lifetime, the French school covered the same ground, as did later generations of criminologists. In particular, Gabriel Tarde, in his 1886 study entitled

La criminalité comparée, demolished Lombroso's argument for the born criminal by employing exacting statistical methods, including a study of four thousand criminals whose supposed atavisms, in particular the shapes of their skulls and brains, were no different from those in the noncriminal population.

As much as the French anthropologists may have disdained Lombroso's methods and conclusions, they were a natural outgrowth of the school founded by Paul Broca. The contemplation of man as a physical object inevitably led to the contemplation of the criminal man as a physical object. Whether such an investigation has merit (and it has yet to be shown that there is a biological basis for criminality), it was natural, if not necessary, that someone should carry it out. The failure to settle the question once and for all is partly a problem of too much information—it would take a lifetime to wade through the writings and studies of the Italian school alone—and the difficulty of getting a handle on the object in question, especially the brain.

It is only fair to point out, as the historian Marvin Wolfgang does, that Lombroso's work predates the advent of modern statistical methods, including linear regression, chi-square analysis, and hypothesis testing—the very tools that legitimized the social sciences. (The subsequent overuse of these tools is another issue.) Lombroso can easily be made to look naive, but he was in fact asking important questions and suggesting (more often than not) reasonable ways of investigating them. According to Wolfgang, most criticisms of Lombroso come from a cursory examination of his work, a problem made more acute because the five editions of *Criminal Man* were never translated into English. Despite his faults, his work "manifested imaginative insight, good intuitive judgment, intellectual honesty, awareness of some of his limitations, attempts to use control groups, and a desire to have his theories tested impartially. Many researchers of today fare little better than this."[20] It is just as easy, Wolfgang claims and then proves, to use selective quotation to portray Lombroso as a sage as it is to unmask him as a fool.

Yet Lombroso was often his own worst enemy. At the fourth International Congress of Criminal Anthropology, held in Geneva in 1896, he was still unwilling to give up on his one cardinal point—the born criminal—and he lashed out at his critics, saying, "What do I care whether others are with me or against me? I believe in the [criminal] type. It is *my* type—I discovered it—I believe in it and always shall."[21]

Shortly thereafter, he became heavily involved in investigations of the occult. At a seance in 1903, he heard the voice of his mother speak from beyond the grave. He instantly converted from a skeptic to a believer. His colleagues worried once more about his reputation, but he could not be moved. He thought that scientific materialism might provide an explanation for telekinesis or communication with the dead, and he set out to find it.

His friends were right to worry. Seen from this vantage point, as a man sitting at a levitating table, listening to disembodied voices and unexplainable rapping, Lombroso is easy to dismiss as a crackpot. But to do so, to reduce him to a single dimension, would be to commit the very sin he was most often accused of.

A plan is now afoot to reconstruct Lombroso's museum as it once was. But because of insufficient interest (and thus insufficient funds), it remains in limbo. With his name no longer a household word, and his theory beyond the pale of current political correctness, Lombroso has become what he would have dreaded, a forgotten man. Meanwhile his face, which is literally kept in a jar by the door of his cluttered time capsule, presides with Ozymandian serenity over the ambiguous remnants of his life's work.

Chapter 8

Séguin

IN THE QUEST to define what is normal, scientists like Cesare Lombroso and Paul Broca naturally focused on the abnormal. If no one could say what an average brain looked like, perhaps they could show what it did *not* look like, and in their search for extreme cases, they sought out not only the brains of geniuses and criminals, but other types of brains that were readily available. Many of them belonged to the mentally impaired and the insane.

Prior to the acceptance of the brain as the organ of the mind, which is to say, prior to the nineteenth century, a mentally retarded person was simply a fool, and there was nothing to be done about it. But that changed when scientific materialism suggested the possibility that an impaired mind signaled a malfunction in the brain-machine. Classifying such malfunctions was regarded as a first step toward understanding them, and the terms that emerged—*idiot* then *imbecile* then *moron* (on an ascending scale of mental age)—reflected the belief that intelligence was, at least in theory, a measurable quality that could be ranked on a linear scale.

This idea got its biggest boost in the 1860s when Francis Galton, Charles Darwin's brilliant but eccentric cousin, wrote a book entitled *Hereditary Genius,* in which he argued that mental ability is not only measurable, but can also be expressed as a single value. Galton chose to use letters for his scale rather than numbers, but the idea was clear enough—everyone is assigned to a rung on the ladder of mental worth, and there they remain. Galton also believed that just as there is such a thing as an average height, there is also an average mental capacity, "and that the deviations from that average—upwards towards genius, and downwards towards stupidity—must follow the same law that governs deviations from all true averages."[1] The law in this case was Adolphe Quetelet's bell-shaped curve, also known as the Gaussian distribution.

Galton's assumption that natural intellectual ability is a single, measurable entity has filtered down to the present in the form of general intelligence, or *g*, the variable presumably measured accurately by IQ tests (which also conform to a bell-shaped distribution). The existence of *g* has still not been accepted by the psychological community, yet Galton not only believed in it, he also proposed to measure it, and not with written exams (they hadn't been invented yet), but through physiological testing. In 1884, at the International Health Exhibition in London, Galton set up an "Anthropometric Laboratory," in which, for only three pence, visitors could be tested in seventeen body measures, including head size (Galton believed in the brain weight theory), reaction time, and sensory acuity. Nine thousand people paid the fee and received a card with their results, and Galton kept a copy for his database.

But in his enthusiasm for measuring, it was never clear, even to Galton himself, just what his measures meant. He once asked his fellow members of the Royal Society to grade their abilities on a twelve-point scale, a task which so frustrated Charles Darwin that he wrote a note in the margin to the effect that "I find it quite impossible to estimate my character by your degrees."[2]

Like Broca and Lombroso, Francis Galton was swept up by the potential of positivist science. Which is to say that he believed in the power of numbers to reveal truths about the human condition.

The problem for Galton was that in relying on the power of statistical averages, he shunned case histories. Which is to say that Galton never produced a compelling discovery narrative that would put a human face on his theory. Lombroso had Vilella, Broca had Tan, and Gall had his school friend with the bulging eyes. Galton, on the other hand, could only point to a bell curve representing a million faceless people.

This chapter deals with the two case histories that personalized the issue of mental retardation and its relation to the brain, and at the same time undermined Galton's contention that everyone occupies a place on a fixed scale of intelligence determined by nature. The first case involved a French boy named Victor, who was found living in the woods, and who was assumed to have grown up there, removed from all human contact. The second was a blind-deaf-mute girl named Laura, who at two years of age found her mind's connection to the outside world reduced to her sense of touch alone. Both Victor and Laura would become famous as subjects of unprecedented experiments that tested a novel idea: that it was possible to overcome mental handicaps with unconventional teaching methods that targeted the five senses. Did feeblemindedness result from a flaw in the brain's structure, as Galton believed, or was it, at least in some cases, the result of a failure of sense impressions to reach the brain?

The story that unfolds below focuses on the investigators who tried to answer this question. One of them, Édouard Séguin, is generally credited with inventing special education. The other, Samuel Gridley Howe, was among other things a pioneer in the education of the blind. Like Gall, Wagner, Broca, and Lombroso, Séguin and Howe were swept up in the revolution brought about by the emerging materialist view of the brain. And although they had no way of knowing the significance of these events, their paths would cross at about the same time that Francis Galton announced his theory of hereditary genius, and with it, his solution to what he called the "eminence gap." The name Galton invented to describe his solution, by the way, was eugenics.

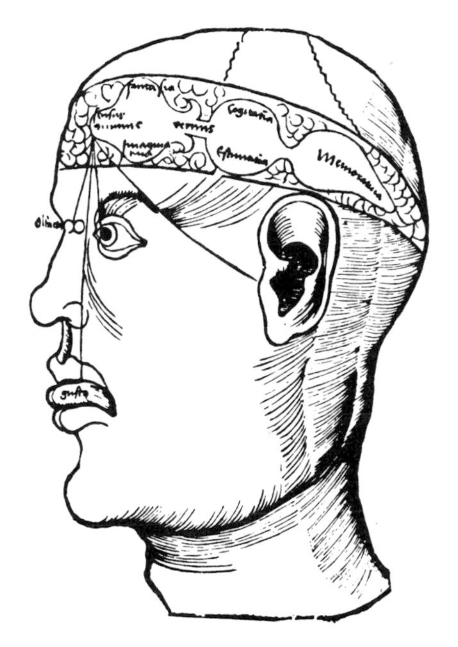

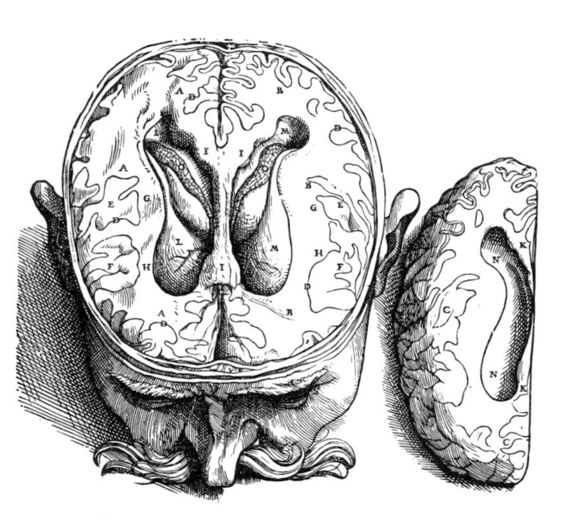

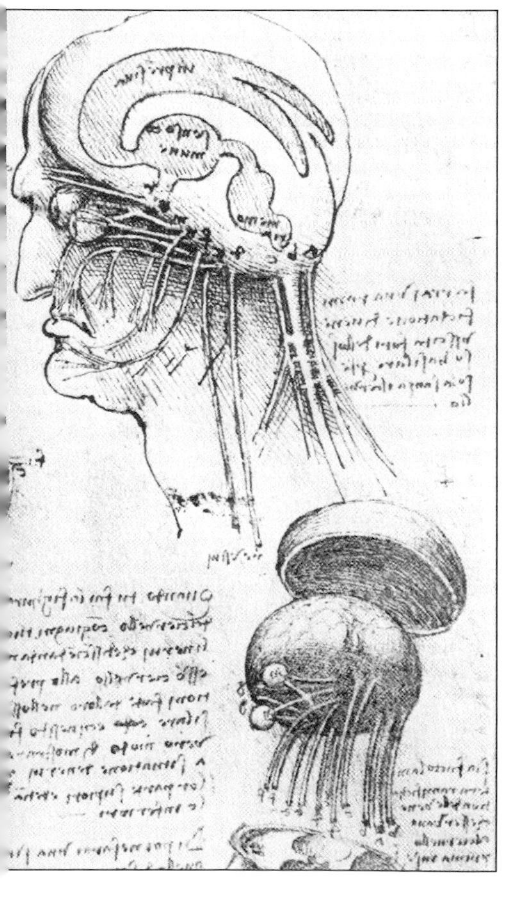

The ventricles, the brain's fluid-filled cavities, shown accurately in Vesalius's woodcut (ABOVE RIGHT) (a vivid illustration of the method of horizontal slices), rather fantastically in a Leonardo da Vinci illustration of 1492 (LEFT), and simplistically in a drawing from Gregor Reisch's encyclopedia of 1504 (ABOVE LEFT). In Reisch's version, the first two ventricles are labeled "fantasy, sensation, imagination"; the middle ventricle "cognition, judgment"; and the last "memory." Leonardo, in his mirror-writing script, uses the terms "imprensiva" (for the confluence of sense impressions), "senso comune" (the seat of judgment), and "memoria."

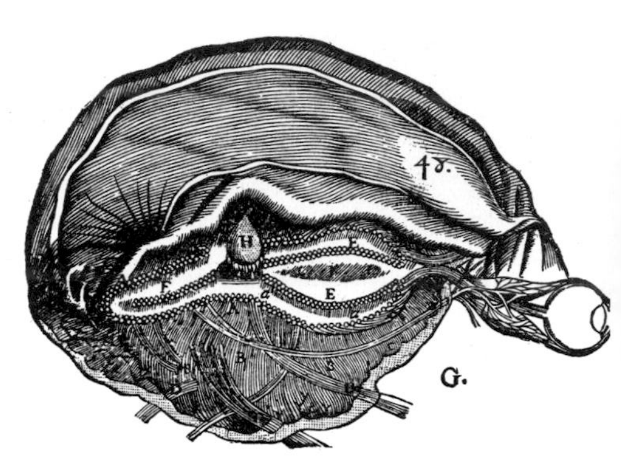

LEFT AND ABOVE: René Descartes's Rube Goldberg conception of the nervous system was a favorite topic for illustrators of early editions of his *Treatise on Man* (this one is from 1664). The brain's central processor, the pineal gland, is labeled H in the unrealistic figure above.

An 1826 print by the Englishman Henry Alken, entitled *Calves' Heads and Brains*, reveals Gall's and Spurzheim's followers as an easy mark for merciless caricaturists. *(By permission of Yale University, Harvey Cushing/John Hay Whitney Medical Library.)*

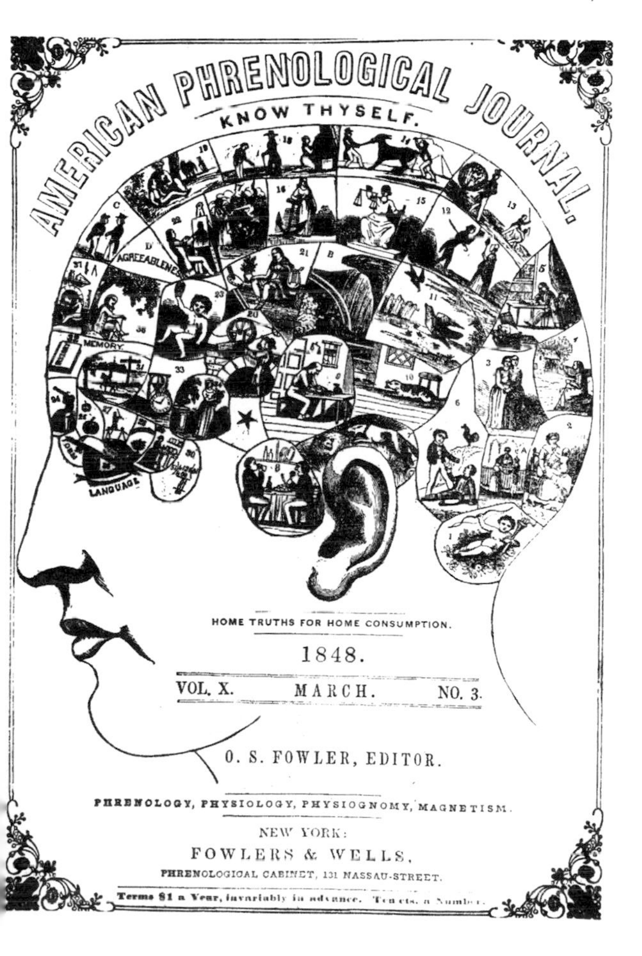
AMERICAN PHRENOLOGICAL JOURNAL.

KNOW THYSELF.

HOME TRUTHS FOR HOME CONSUMPTION.

1848.

VOL. X. MARCH. NO. 3.

O. S. FOWLER, EDITOR.

PHRENOLOGY, PHYSIOLOGY, PHYSIOGNOMY, MAGNETISM.

NEW YORK:
FOWLERS & WELLS,
PHRENOLOGICAL CABINET, 131 NASSAU-STREET.

Terms $1 a Year, invariably in advance. Ten cts. a Number.

ABOVE: The Phrenological Cabinet of Fowler and Wells, after the firm moved from Nassau Street to 308 Broadway. From the *New York Illustrated News*, February 18, 1860.

LEFT: An 1848 title page from the Fowlers' *American Phrenological Journal*, vol. X, no. 3, the year before Walt Whitman decided to have his bumps read. Whitman was pleased to discover that "adhesiveness" (region number 3, soon relabeled "friendship") was one of his most outstanding qualities.

ABOVE LEFT: Three heads from Gall's collection (two of them casts of Gall himself) on a storage room worktable at Paris's Musée de l'Homme. ABOVE RIGHT: The cabinet that inspired Carl Sagan's *Broca's Brain*, containing the remnants of the Société Mutuelle d'Autopsie. IMMEDIATELY ABOVE: A cabinet at the Musée Dupuytren, containing the brains of Broca's famous patients, Leborgne (at left) and Lelong (left-center). A portrait of Broca can be seen lurking behind the skulls. *(Photos by the author.)*

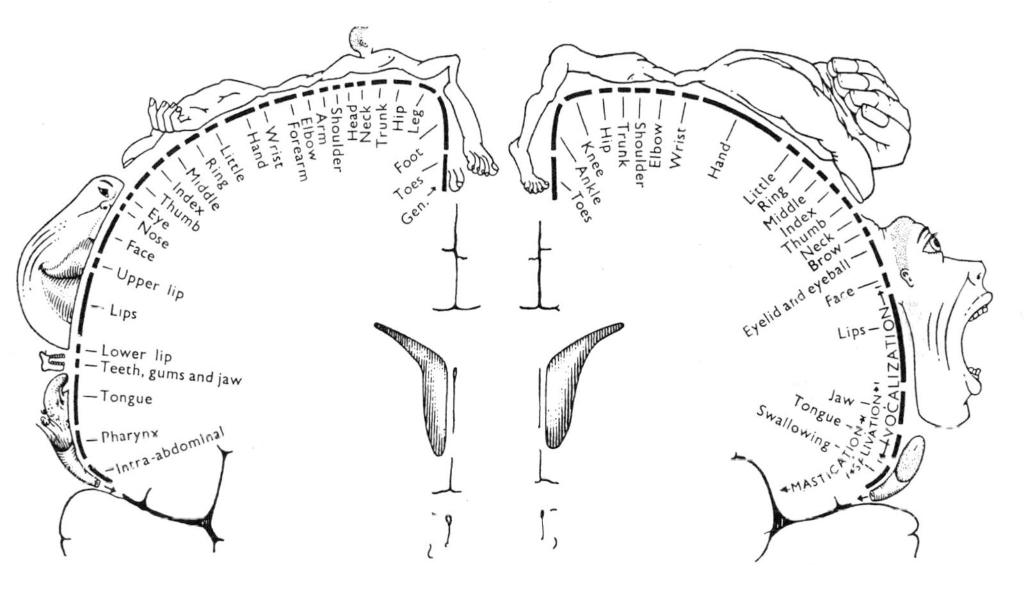

Phrenology was partly vindicated in the 1870s with the discovery and mapping of the sensorimotor cortex. Two classic illustrations from Wilder Penfield and Theodore Rasmussen's *The Cerebral Cortex of Man* (1950) vividly demonstrate the proportional structure of the motor and the sensory cortexes. (The brain is seen here in a coronal section, which is to say, sliced over the crown from ear to ear, and viewed from the front; the deep infolding is the Sylvian fissure.) At the top is the so-called motor homunculus; below is the sensory homunculus. *(From* The Cerebral Cortex of Man, *by Wilder Penfield and Theodore Rasmussen, © 1950 Macmillan. Reprinted by permission of The Gale Group.)*

The Wilder Brain Collection as it looked a century ago. Note the remarkably smooth brain of Chauncey Wright (labeled "Philosopher and Mathematician"), second from the left on the bottom shelf. *(Brains of Educated Orderly Persons. Burt Green Wilder Papers, #14/26/95. Courtesy of the Division of Rare and Manuscript Collections, Cornell University Library.)*

The anatomist James Papez, the man who removed and studied the brain of Burt Green Wilder, poses with the Cornell collection in the 1940s. *(Professor James Wenceslaus Papez with brain collection, inspecting a specimen. Faculty Biography Collection. Courtesy of the Division of Rare and Manuscript Collections, Cornell University Library.)*

LEFT: Gauss's brain, death portrait, and a plaster cast of the interior of his skullcap, in front of the cabinet containing the brains of Dirichlet, Fuchs, and Hermann. On top of the cabinet is Gauss's likeness on the ten-deutsche-mark note. *(Photo by the author.)*

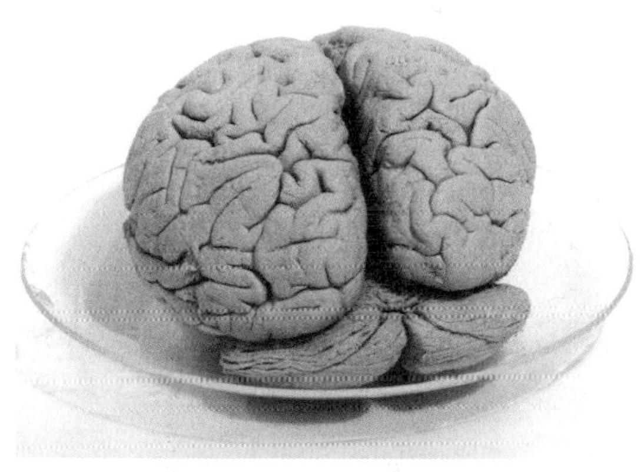

RIGHT: The brain of Gauss, ready for its MRI, in 1999.

BELOW: The resulting brain scan, revealing a healthy specimen with no signs of senile atrophy. *(Wittmann et al., "Magnestresonanz-Tomografie des Gehirns von Carl Friedrich Gauss."* Gauss-Gesellschaft E.V. Göttingen Mitteilingen, *no. 36 (1999), pp. 9-19. Reprinted by permission of Axel Wittmann on behalf of the Gauss Society.)*

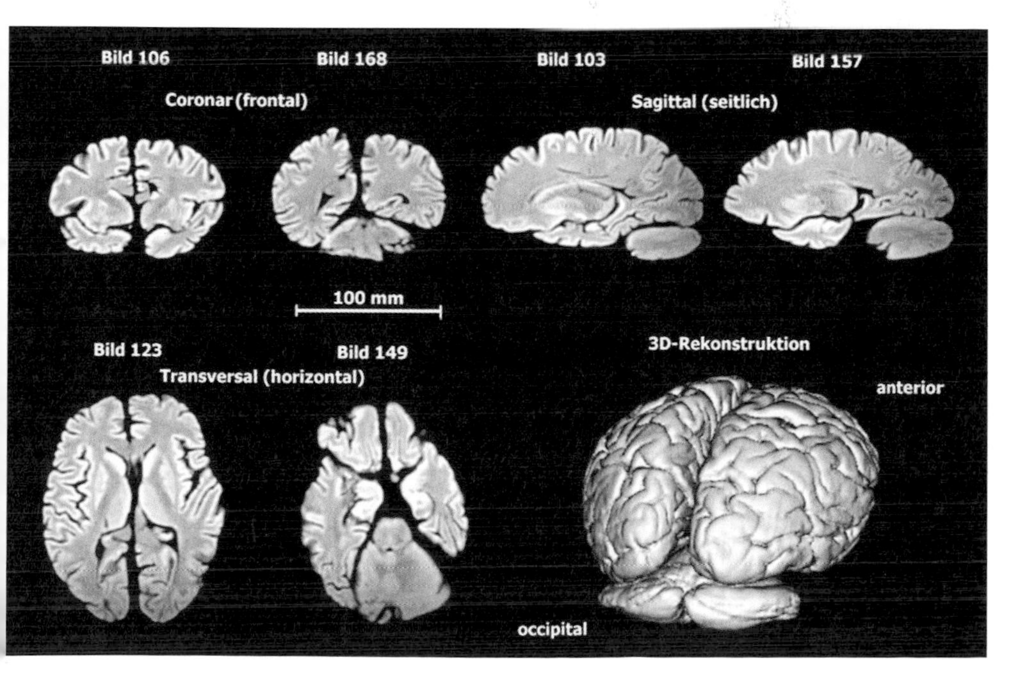

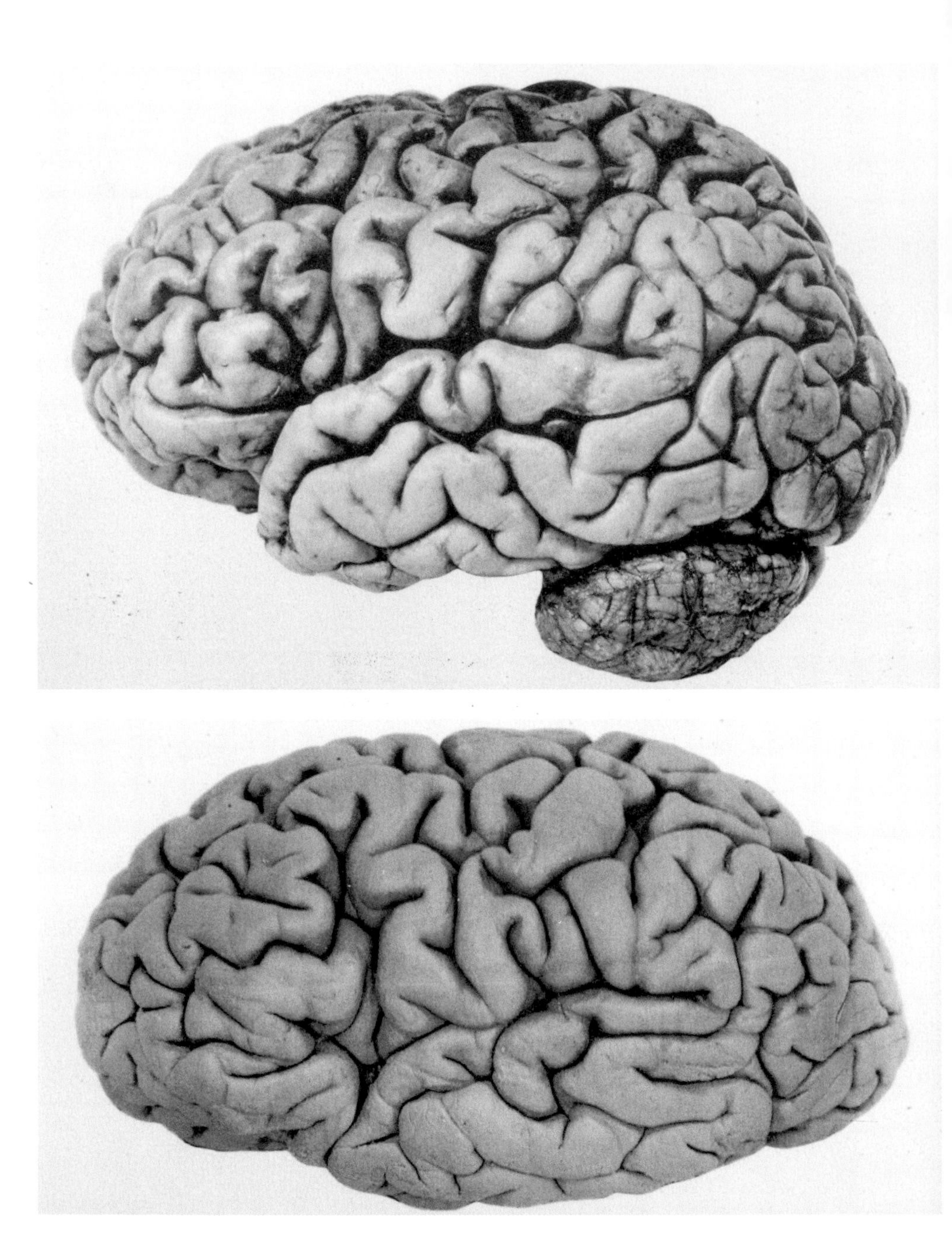

The new phrenology, as practiced by Edward Anthony Spitzka and Sandra Witelson, is a game anyone can play. In photos of the left hemispheres of the Swedish astronomer Hugo Gylden (top) and the Russian mathematician Sonya Kovalevsky (bottom), prepared by the anatomist Gustav Retzius (who refrained from reading meanings into the fissures), Spitzka saw "a struggle for expansion" in Gylden's parietal region, and suggested a possible connection to his mathematical talents. He was less effusive in praise of Kovalevsky's brain (if not of all women's brains), despite noting similarities to Gylden's. *(By permission of Yale University, Harvey Cushing/John Hay Whitney Medical Library.)*

ÉDOUARD SÉGUIN'S STORY begins in 1799, when three hunters, walking through the woods on the outskirts of the village of Aveyron in southern France, came upon a filthy and naked boy who seemed to be foraging for roots and nuts. The boy looked and behaved just like a wild animal, and he climbed a tree when the men approached. With little effort, they coaxed him down and brought him to a local widow who tried to care for him. But the boy longed for open spaces and soon managed to escape. An unusually cold winter came and went, and in the spring the boy was spotted and captured again, and this time kept under closer guard.

The boy appeared to be about ten or eleven years old, and he could not speak. He loped rather than walked. For a time he was placed on public display and promoted as a "specimen of primitive humanity."[3] While some townspeople thought he had probably been abandoned by a local family, others claimed to have caught glimpses of him in the woods up to five years earlier. If true, the boy would have grown up outside of any human contact in a hypothetical state of nature.

News of the boy eventually reached Paris, and he was sent for by the medical authorities. In 1801, the director of the Bicêtre asylum, Philippe Pinel, made a thorough examination and pronounced the boy to be an incurable idiot. Such was the opinion at the time about the severely mentally retarded. They were written off as lost causes. But a colleague of Pinel's named Jean-Marc Itard came to a different conclusion. Itard thought that the boy had been abandoned, that he was merely unschooled, and he offered to take on the task of educating him. Itard named the boy Victor (the very name, perhaps not coincidentally, that Mary Shelley would choose for the protagonist of her famous novel). He became more widely known, however, as the Wild Boy of Aveyron.

The case of Victor, one of only a few documented instances of so-called feral children, provided a unique opportunity to test popular theories of the mind. Itard, for example, echoing John Locke, believed that the total content of the human mind is built up of experience, and he devised a program of rehabilitation based on reawakening Victor's senses.[4] Although Victor never did learn to speak more than a few words, his socialization progressed to the point where he could understand simple language, read basic phrases, and behave outwardly like a normal child. But after five years of struggle, Itard gave up his project as a failure and placed Victor in a foster home. The boy, who would live for another thirty years, was soon forgotten, but Itard's experiment was not.

In 1837, near the end of Itard's career, another retarded boy arrived at the Bicêtre, and Itard faced a dilemma. Feeling too old for the job, he hesitated to repeat his experiment, and turned the case over to one of his students, along with all of his files on the wild boy Victor. The student was Édouard Séguin.

At the time, Séguin was a man of big ideas but very little practical experience. He had just started working at the Bicêtre (for no pay), and he had yet to find his métier. But experienced or not, Séguin was extremely perceptive, and as he reviewed the files on the Wild Boy, he became convinced that his teacher had made a crucial mistake. Victor, Séguin realized, was not a physically normal child who merely lacked schooling. He was, in fact, an idiot—a child whose brain had, for whatever reason, not developed properly. Moreover, Pinel had also been wrong in saying that Victor was incurable. The boy had made significant progress, which led Séguin to believe that any retarded child could do as well, if not better, with an earlier start. The trick, he decided, was to retrain the brain from scratch.

In the 1830s, idiocy was defined as a failure to attain "moral maturity," a failure usually attributed to brain diseases that, after Morel, would be blamed on degeneration. (A "lunatic," by contrast, was someone who had attained moral maturity, and then had lost it.) Séguin refused to accept this verdict, and he proposed a radically new theory of idiocy along with a program to cure it, or at least mitigate its effects. The idiot child, he said, did not suffer from a disease of the brain as much as a lack of physical development. Some circumstance, either prenatal or postnatal, usually malnutrition, had prevented the child's brain from growing at the proper rate. In order to correct this, Séguin devised a series of exercises aimed at awakening the brain through the five senses, and he supplemented these with a vigorous program of physical education and proper diet. Whether the underlying theory was true or not, the program worked.

ÉDOUARD SÉGUIN HAD an unusual résumé. He was born in 1812 in Clamecy, a small town on the river Seine in France's Burgundy region. Upon arriving in Paris in 1832, he studied medicine at the Salpetrière mental hospital, but was discharged in 1834 after a scandal, and seems not to have completed his degree. Three years later, he met Itard, and began working for him at the Bicêtre asylum, still without an M.D. (Itard himself never studied medicine formally, but learned it on the job during the French Revolution.) It was while working with François Leuret at the Bicêtre that Séguin founded a school for retarded children at the Hospice des Incurables, partly at Itard's suggestion.

Teaching for no pay, Séguin supported himself as an art critic, and he joined a select circle of writers that included Victor Hugo and Pierre Flourens. Like Itard, he was a freethinker and an atheist. Like Condillac and Cabanis, he was a scientific materialist. But he drew most of his inspiration from the works of Claude-Henri de Rouvray de Saint-Simon, whose socialist philosophy, best expressed in his 1825 book *The New Christianism*, had united a generation of scientifically minded social reformers, including Auguste Comte.

"The whole of society," declared Saint-Simon, "ought to strive toward the amelioration of the moral and physical existence of the poorest class."[5] Séguin rose to the challenge, and his school soon gained wide attention and praise. After only six years of operation, the French Academy of Science declared it a success and a model for other schools. Séguin, they announced, had solved a long-standing problem: idiotic children could indeed be educated using scientific methods. Within a decade, the eccentric teacher had emerged as the leading authority in his field. His 1846 textbook—*The Moral Treatment, Health, and Education of Idiots and Other Backward Children*—would become the standard reference work on the subject. His "physiological method" of educating the senses later became the inspiration for the Montessori method, and it remains the basis of special education to this day.

During the 1840s, teachers, reformers, and philanthropists from around the world flocked to Séguin's Hospice des Incurables, and spread his gospel by opening schools of their own. One of them, Horace Mann, the "father of American public education," toured the school in 1842. Another influential visitor, George Sumner (brother of the Massachusetts senator and strident abolitionist Charles Sumner), wrote a series of enthusiastic letters describing the miracles he saw in Séguin's classroom.

But the recipient of these letters, a world-famous teacher of the blind named Samuel Gridley Howe, was not won over by them. Howe was troubled by a glaring deficiency in Séguin's program—its lack of a religious foundation. Unlike his friends Mann and Sumner, Howe remained skeptical, and regarded the so-called French School as a challenge rather than an inspiration.

AS THE FOUNDER of special education in America, Samuel Gridley Howe was Edouard Séguin's American counterpart, although in most respects—temperament, religious beliefs, and educational philosophy—the two men were poles apart.

Howe was born in Boston in 1801. After graduating from Harvard

Medical School in 1824, he decided to style himself as a romantic hero in the mold of Lord Byron (albeit without the literary talent), and he set off to fight for Greek independence. When he returned after six years of daring escapades, he set out to change the world.

Howe thrived on challenges. His personal motto was: "Obstacles are things to be overcome," but he also yearned for respectability. So when his friends offered him the chance to direct a new school for the blind they planned to build in Boston, he took it. To prepare himself, he traveled to Europe to visit similar schools. (He could not manage to stay out of trouble. While visiting Prussia he was jailed for six weeks for bringing aid to Polish refugees.) Upon his return in 1831, Howe opened his school—the Perkins Institute for the Blind—with six students.

Samuel Howe never did anything strictly for its own sake or for personal gain. He was a true idealist, a moral crusader who believed that his every act should have a higher purpose. By teaching the blind, by aiding refugees, or by running guns for the Greeks, he was doing God's work, and his success was the proof of it. Obstacles, for Howe, were not merely things to be overcome, but to be sought out. They had to be large enough to be worthy of his time, and so improbable, if not impossible, that his eventual success would stand as incontestable evidence of a divine plan. When he opened his modest school, the real challenge was not just to teach but to validate his mission in life. What he needed, of course, was his own Wild Boy of Aveyron, a success story in which he could play the role of miracle worker. In 1837, he found it in the person of a little girl named Laura.[6]

Laura Bridgman, a seven-year-old blind deaf-mute, had grown up on a farm outside of Hanover, New Hampshire. She was only two years old when scarlet fever ravaged the family, killing her two sisters and robbing Laura of all sight in her left eye and most of it in her right. It also destroyed her hearing and stunted her senses of smell and taste. To make matters worse, what little sight remained to her was snuffed out in a horrible accident two years later when she walked into the spindle on her mother's spinning wheel. The girl learned to communicate as best she could using a primitive set of gestures and noises, but after these catastrophic events, her developing mind was suddenly cut off from the world around her.

Howe first learned about Laura from a friend, a Dartmouth College professor, who thought that the girl would be a perfect candidate for the Perkins School. Howe agreed, although he had a larger goal in mind. Through Laura, Howe saw a chance to demonstrate a theory of mind based on a corruption of Caspar Spurzheim's phrenology. Howe believed in the existence of innate faculties, but refused to accept the materialistic implications of Gall's organology. He was convinced that God predetermined the character of each person through the shape of his or her brain, and that each mind was endowed at birth with a faculty of veneration. If he could show that Laura's faculties needed merely to be kindled by the right kind of input, she would stand as living rebuke to scientific materialism. If he could show that with no religious education of any kind, Laura would find God on her own, he would validate his personal vision. Through Laura Bridgman, the perfect test case, Howe set out to prove the existence of the soul.

It should come as no surprise that in his memoirs, Howe recounts a eureka moment, one in which Laura finally made the connection between the raised letters of the books that he had made for her and the objects he placed in her grasp. Once this happened, Laura's language faculty kicked in at an alarming rate, and she became a prolific writer, and such an avid reader that she often forgot to eat.

Laura Bridgman, it turned out, was not feebleminded. She was, in fact, a literary prodigy. She wrote poetry, childrens books, and kept a journal, and as the first blind deaf-mute to be taught to read and write, her reputation spread quickly. During the next six years, teacher and pupil became international celebrities on the lecture circuit. Charles Dickens, who met her on his trip to America in 1842, ranked Laura second to Niagara Falls as his most memorable encounter. But for Howe, the experiment was not complete. He had carefully shielded Laura from religious ideas and influences, biding his time until her faculty of veneration would bloom on its own. Too late, he learned that he could not keep the outside world at bay.

Howe's great experiment effectively ended in 1843 when he married Julia Ward (who would become even more famous than her husband by penning the lyrics to the "Battle Hymn of the Republic"). They honeymooned in Europe, leaving the thirteen-year-old Laura in the care of a tutor, who took the opportunity to convert the impressionable girl to Christianity. Howe returned to find his project in ruins. He despondently noted, "I hardly recognized the Laura I had known," and his interest in her never recovered. He now had to look for another way to prove his theory.

As it happened, Howe already had another project in the works. In 1839, while Laura was just beginning to show signs of literary skill, a blind, retarded child was brought to him for instruction. Howe made some progress with the boy, mostly through sheer persistence. Like Édouard Séguin, he was unwilling to concede the hopelessness of educating any child. But as he expanded the project, he found that the retarded students did not mix well with the blind ones. (Laura in particular dreaded her encounters with them. "I would be so happy," she once wrote, "if they could prescribe the Idiots not to have our rooms.") So Howe, with Sumner's encouragement, raised more money and built another school, calling it "a beautiful example of practical Christianity; a temple in which acceptable service is done to God."

Howe employed the ablest teachers he could find, hoping to effect a divine awakening in the minds of his new students. His first hire, James B. Richards, was a teacher of "unrulies" from New York, who believed that "the lowest idiot and criminal could be reached and humanized if only enough love were spent on him." Howe encouraged Richards to visit Séguin's school in Paris, which he did in 1849, although the experience left him unimpressed. "There was little or nothing to be learned there," he wrote back to Howe, "that would not have been arrived at here by any intelligent person."[7]

Séguin, of course, steered clear of religious instruction. His school was founded on scientific principles alone, and looked to the blank slate as the proper model of the human mind. His physiological method may have impressed Howe, but his atheism did not. Even so, when James B. Richards left Howe's school in 1852 to take over a similar school that had been founded in Philadelphia, Samuel Gridley Howe took the unusual step of hiring Édouard Séguin to replace him.

IN 1850, OUT of fear of reprisals for his support of the failed revolution of 1848, Séguin had packed up his small family and emigrated to America. He established a modest medical practice in Cincinnati, but soon felt out of his element, and was so desperate to return to education that he gladly accepted Howe's offer. His reputation was by then legendary, but his new employer had no idea what he was getting.

Over the next two decades, Séguin would teach or consult at most of the schools for retarded children in the United States. These arrangements never went smoothly, partly because Séguin was not a team player. He was subject to fits and spells of "depressing disease," and as a freethinker he found himself out of step with American culture. In particular, he did not share Howe's enthusiasm for phrenology and spirituality, and the two men failed to get along from the start. After only three months on the job, Séguin left Howe's school in order to take a job at the Syracuse State School. Soon afterward, he accepted a similar position at the Pennsylvania Training School in Germantown, just outside of Philadelphia. But the school's director, Howe's former assistant James B. Richards, had even less luck with the quirky teacher. In what would become a familiar routine, Séguin stayed for three months, taught the staff his physiological method, began to feel uncomfortable with office politics, and left. He would do the same at a dozen schools over the ensuing decade. As one of his dearest friends would later remark, Séguin "was instrumental in establishing schools for idiots in various parts of our country, which are living, working monuments of his life-long labors."[8] They were also monuments to Séguin's notorious inability to work for anyone but himself.

Like all of the schools Séguin visited, the Pennsylvania Training School adopted his methods wholeheartedly, regarding him as a patron saint. Founded in 1852 with fewer than twenty students, its success soon forced it to move to larger quarters. In 1857, the school acquired a new director, Isaac Newton Kerlin, a recent graduate of the University of Pennsylvania School of Medicine, who would eventually succeed Séguin as the country's preeminent educator of feebleminded children. He would also, in time, be won over by Francis Galton's theory of hereditary ability. Which is to say that Isaac Kerlin would take an interest in eminence and genius, while introducing eugenic measures into American schools for the retarded.

ISAAC KERLIN BEGAN his career as a disciple of Édouard Séguin. He was a tireless educator with a talent for administration, essentially a born leader. After a stint in the Union Army during the Civil War, he presided over the institutionalization of special education in America, which saw a scattering of small schools founded or inspired by Séguin give way to planned communities on the scale of industrial villages.

Kerlin's reign began officially in 1876, when a select group of school directors, including Séguin, founded the Association of Medical Officers of American Institutions for Idiotic and Feeble-Minded Persons (AMO). Séguin was elected president, and Kerlin volunteered his services as the group's secretary-treasurer, a position that afforded him the real power in the organization. Because the group met regularly at the Pennsylvania Training School (now relocated fifteen miles outside of Philadelphia), and because Séguin had no patience for committee work, Kerlin dominated the meetings, and he set the organization's agenda for the next two decades. (Séguin soon distanced himself from the group, which had already begun to distance itself from his methods.) Within ten years of its founding, the AMO had replaced Séguin's vision of individualized care with the so-called "colony plan," a network of rural villages for the feebleminded in which the residents were employed (without pay) in agricultural work or handicrafts.

In addition to being a relentless organizer, Kerlin was also a mercurial and unpredictable man who could show infinite patience with an individual child, and no patience at all with children in general. He was apt to fly off the handle in response to a careless remark from a subordinate. He even kept a journal in which he logged his secretary's spelling mistakes. His assistant and eventual successor, Dr. Martin Barr, once remarked that Kerlin brought to his work "a power that was almost mesmeric, and that controlled absolutely all who came within its influence. Children, attendants, officers, friends, visitors, the most careless, the most indifferent, testified to this undefined 'something' to which all yielded and that made itself felt equally within institutional walls as in legislative halls."[9]

His grip on the everyday workings of the school may have kept his staff on edge, but it had its positive side. Kerlin saw to it that no employee ever abused a child. "Attendants and others are positively prohibited carrying switches, sticks, canes, etc.," he decreed. "The position of attendant is a sacred one."[10] But as kind and patient as he might have been with individual children, he increasingly came to view the feebleminded population as a whole with suspicion and distrust.

Like most educators of the period, Kerlin adhered to Bénédict Morel's degeneration theory. "Tendencies to congenital cerebral disease of offspring," he wrote, "are established through practices and vices which lower the morale, impair the strength, and vitiate the blood of ancestors and parents."[11] Although followers of Morel believed that degeneration led to sterility, that it bred itself out of the line of descent after a few generations, Kerlin eventually recognized the need to speed up the process. According to historian Nicole Hahn Rafter, he "came to regard eugenics as the chief weapon in the war against mental defect and institutionalization as the best method of achieving reproductive control."[12] And he was not alone.

The transformation of the training school at Elwyn during Kerlin's thirty-five-year tenure from a small, hands-on rehabilitative center into an agricultural and eugenic colony with over a thousand inmates was mostly a matter of economic necessity. In the early stages of special education, most of the students who attended schools like Samuel Howe's were orphans or abandoned children. By the time they reached maturity, the school was the only home they had known. Even after years of training, many could not survive on their own, and they faced the grim prospect of living in almshouses, or worse. The only humane alternative was to keep them on as workers or inmates. This is how the schools for the retarded evolved into institutes—part training facility, part asylum. The Pennsylvania Training School, for example, was renamed the Elwyn Institute in 1871, when it was split into a school division and a custodial division. Its growth had been phenomenal. In 1868, it had housed 150 students; by 1877, the number had risen to 300; and, by the late 1880s, it had swelled to over 1,000, most of whom lived and worked in its custodial division. Nationwide, the number of feebleminded residents in state-run institutions was about 40,000.[13]

The school directors brought part of this problem on themselves. More students meant more money from the states, and more pressure to expand. In response, Kerlin and his associates began to loosen their definition of feeblemindedness. Just as Cesare Lombroso resorted to degeneration as a way to expand his class of criminal types, the school directors spread their net wide when they created the "congenital moral imbecile," a category that could include almost anyone, although it was most often applied to clever boys who had run afoul of the law. (It even applied to those who were *likely* to become criminals.) Kerlin explained that even though congenital moral imbeciles were often "precocious in the power to acquire school learning," they "should be withdrawn from the community before they reach crime age."[14] So he withdrew them.

By 1903, the connection between feeblemindedness and criminality was so firmly entrenched that Walter Fernald, director of the Massachusetts School for the Feeble-Minded (the school founded by Howe, and later renamed for Fernald), could confidently proclaim that "a large proportion of our criminals, inebriates and prostitutes are really congenital imbeciles who have been allowed to grow up without any attempt being made to improve or discipline them."[15] On the positive side, as one school director boasted, moral imbeciles "made first class foremen."

FOR MOST OF his long career, Francis Galton advocated something that is now called positive eugenics. His solution to the "eminence gap" focused solely on the high end of the achievement scale, on "increasing the productivity of the best stock." In other words, Galton recommended intermarriage within the intellectual elite, and encouraged large families, arguing that his symmetric bell curve would quickly become skewed toward the upper reaches if this was done even on a modest scale. "If a twentieth part of the cost and pains were spent in measures for the improvement of the human race that is spent on the improvement of the breed of horses and cattle," he wrote, "what a galaxy of genius we might not create! We might introduce prophets and high priests of civilization into the world as surely as we can propagate idiots by mating *crétins*."[16] Only very late in his career did Galton shift his priorities and begin to promote negative eugenics—the restriction of the breeding of undesirables. By that time, it had been a standard practice in the United States for a quarter century.

The practice of isolating feebleminded women of childbearing age began in 1878, when the state of New York established the first asylum exclusively for adult, mentally retarded women. Although retardation was not proven to be a heritable trait, the asylum intended to "guard against the increase of this class."[17]

At first, Isaac Kerlin disapproved of the practice, but changed his mind a year later when he read a book on eugenics by a Scottish physician. Kerlin would not live long enough to preside over the custodial mentality that led to the rounding up of drunks, the poor, and tramps into a system that made routine use of sterilization, and that replaced a culture of rehabilitation with one of conformity and obedience. (Most of that would take place in the early twentieth century under his successor, Martin Barr.) But he did initiate another practice with equally sinister overtones. While instituting eugenic measures in his school, he also began to collect brains of retarded children.

In 1884, Kerlin hired Alfred W. Wilmarth, a graduate of the University of Pennsylvania School of Medicine, to serve as his medical director, and he placed him in charge of a new pathology laboratory at Elwyn. Wilmarth performed autopsies on hundreds of children who died at the school, and preserved their brains for study. At that time, the mortality rate for retarded children between the ages of five and ten was 50 percent; for ten- to fifteen-year-olds the rate dropped to 34 percent; but for fifteen- to twenty-year-olds it rose to 45 percent. The leading cause of death was infectious disease, notably rubella and tuberculosis, which kept Wilmarth well supplied with material.

In an 1890 study of one hundred brains of feebleminded children—thirty-four from the school department, and sixty-six from the custodial department—Wilmarth found that the majority suffered some form of cerebral disease, usually multiple sclerosis, an affliction that attacks the myelinated nerve fibers and often leads to paralysis.

He also noted several cases involving microcephalic children whose brains had either not developed, or had been prevented from developing by the early closure of the skull's sutures. Five cases involved what he referred to as "mental defect," by which he meant Down's syndrome. He concluded, rather grimly, that most feebleminded children would never attain full mental development, perhaps an obvious point, but also a cruel one when combined with his recommendation that "the physician should be more cautious in giving customary diagnosis of 'Arrested development that special training will remove,' thereby arousing hopes in the minds of the parents that can only result in disappointment."[18] Wilmarth seems to have viewed the matter in black and white, as though anything less than a complete recovery was not worth the effort.

Wilmarth published his results in the *Proceedings* of the AMO, which was hardly a scientific journal. But then Wilmarth's studies were not very scientific. With no references, notes, diagrams, or statistical analyses, they were little more than "observations." Wilmarth, one of the earliest advocates of forced sterilization of women, made his only real contribution to science when he began to send the brains of his feebleminded children to the Wistar Institute of Anatomy and Biology in Philadelphia. To his credit, he eventually volunteered to send his own.

Édouard Séguin, Isaac Kerlin, and Alfred Wilmarth had this much in common: a willingness to give their own bodies to the cause. When Séguin died in 1880, his was one of the first brains in America to be preserved. Although it was done at his own request, the brain was not studied for another twenty years out of deference to his son, the neurologist Edward Constant Séguin. When the son died in 1898, his brain was also removed, leading to the first comparative study of father-and-son brains. In the early 1890s, Kerlin and Wilmarth joined a brain-donation society that was based at their alma mater, the University of Pennsylvania. Wilmarth would harvest the brains of Kerlin and his wife, Harriet, in 1893.

IT WAS INEVITABLE that the study of mental retardation would come within the purview of anatomy, that the quest to understand cases like Laura Bridgman's would end up on the dissecting table. By the 1880s, medical science enjoyed such high esteem that the thought of donating one's brain, or appropriating someone else's, carried little or no stigma.

In 1879, Laura agreed to undergo a physiological evaluation by the pioneering experimental psychologist Granville Stanley Hall. The resulting study, published in the psychology journal *Mind* later that year, told the story of her illness, her education, and her adult life. It then summarized the results of a physical examination as well as a thorough psychological profile. Hall belonged to a new school of psychologists that emphasized the importance of physical measurements. It is not surprising that he wanted to carry the examination further. At one point he remarked, rather ominously, that a "*post-mortem* examination" would be "extremely desirable," and Hall would get his wish.[19]

When she died ten years later, Laura's brain was removed at Hall's request (with the eyeballs attached) and preserved for study. Hall tapped a former student, the neurophysiologist Henry Herbert Donaldson, to examine the specimen. In a lengthy report that ran in three parts in the *American Journal of Psychology,* Donaldson noted a few peculiarities in the brain, principally a failure of the frontal and temporal lobes to close over the insula. (Essentially, the Sylvian fissure on each side was so wide as to reveal the usually hidden lobe that lies beneath it.) Although not a rare feature in brains, it was unusual enough, and extreme enough in Laura's case, to suggest a possible connection with her disabilities. Still, Donaldson would not venture to say what that connection might have been. Atrophy of the optic nerves, which he also noted, was to be expected, but in the brain itself, specifically in the speech and auditory centers, he did not see anything suggestive of Laura's impairments. He noted that the cortex was, as a whole, not very well developed, its thickness being well below the average. But he did not attempt to phrenologize the specimen.

In hindsight, it comes as no surprise that brains belonging to those in the lowest and highest rungs of Francis Galton's ladder of mental ability failed to confirm his model. No single measure of them, neither weight nor volume nor fissural development nor ratios of lobes, corroborated Galton's theory and fitted a bell-shaped distribution. Galton had focused on extremes, on extraordinary genius and hopeless idiocy. Yet he failed to explain how to assign a criminal, a lunatic, a Wild Boy, a blind girl, or even a congenital mental imbecile to his or her proper grade on his scale of worth. Nor did he consider the possibility that mental ability might lie dormant, requiring only the intervention of a caring teacher to make its presence known.

Laura Bridgman lived long enough to have her fame eclipsed by that of Helen Keller, just as Samuel Howe's was eclipsed by that of Annie Sullivan, Keller's famous tutor, who had trained at the Perkins School from 1880 to 1886. (Howe never met her. He died in 1876.)

Laura's brain no longer exists. Unlike the brains of eminent men collected during her era, it was subjected to a complete investigation, requiring dissection and eventual destruction. Before cutting into it, Henry Donaldson photographed the brain, then commissioned a local artist to make an exact rendering in clay, from which he made plaster casts. One of them is still kept in a display case at the Perkins Institute for the Blind in Watertown, Massachusetts, as a reminder of an extraordinary mind that was once so cut off from the outside world, and needed only to be awakened.

Chapter 9

Guiteau

ON THE MORNING of July 2, 1881, just six months into his first term as president of the United States, James A. Garfield was gunned down as he arrived at the Baltimore and Potomac rail station in Washington, D.C., en route to commencement ceremonies at his alma mater, Williams College. As Garfield and his entourage strolled through the nearly deserted great hall, a bearded man walked up behind him, drew a pistol, fired two shots, and walked away. One shot hit Garfield in the arm, the other in the back. A physician at the scene tried to reassure the president, but Garfield, who had seen plenty of bloodshed during the Civil War, replied, "I thank you, Doctor, but I am a dead man."[1]

He wasn't, at least not yet, and there is some speculation that his wound might have healed on its own. Instead, over the next six weeks, a succession of surgeons tried to dig out the bullet. Due to this constant probing, some say, Garfield died on September 19 of a burst aneurysm. The gunman, an eccentric attorney named Charles Guiteau, stood trial for murder two months later. The public was out for blood, and cared little that by almost any standard of that time (and by every standard in use today), Guiteau was out of his mind before, during, and after the shooting.

As he perused the newspapers in the weeks that followed, a New York neurologist named Edward Charles Spitzka came to the same conclusion, and the impression only hardened when he saw Guiteau's picture. The asymmetrical face, the lopsided smile, the strangely formed head could only have belonged to a lunatic. Guiteau's statements to the press merely confirmed the diagnosis. He claimed to have been instrumental in getting Garfield elected, and was then rudely denied an appointment in the new administration. The man was clearly deluded.

At the time, Spitzka was only twenty-nine years old, but already a nationally recognized expert on insanity. As the trial of Guiteau drew near, both the prosecution and the defense tried to purchase Spitzka's testimony, but he declined all offers. Almost a dozen expert witnesses would eventually be called to testify, and Spitzka did not intend to be one of them. Only at the last moment was he served with an attachment—essentially a subpoena—which forced him to appear for the defense. His performance would be the high point of the trial.

By all accounts, Edward Spitzka was an irascible man. He spoke with complete self-assurance, without pause or hesitation, and was cowed by no one. A fellow neurologist once called him "a red-headed, hot-headed man," adding that he "made us sit tight in our chairs unless we had something important to say."[2] Despite his age, he was also one of the world's foremost authorities on the human brain. Just a year earlier, he had removed and preserved the brain of his good friend Édouard Séguin, intending to conduct a detailed study. At the trial of Charles Guiteau, a dream team of prosecutors would take turns trying to tear Spitzka's reputation to shreds. In the end, they would regret having asked him anything at all.

EDWARD CHARLES SPITZKA was born in New York City in 1852. His father, a German watchmaker named Charles Spitzka, had come to America on the same tide of European unrest that had brought Édouard Séguin. The precocious boy would attend Thomas Hunter's celebrated P.S. 35, then City College, and finally the Medical School of the University of New York, before heading off to Europe to finish his education.

Spitzka belonged to the last generation that was obliged, because of the backwardness of American professional schools, to study medical specialties in Europe's great universities. In Leipzig he worked under Hermann Wagner, the son of Rudolf Wagner; in Vienna he learned embryology, neurology, and psychiatry from the great Theodor Meynert, one of the pioneers in the neurophysiology of the human brain. When he returned to New York, he brought the latest European ideas with him, in particular a materialistic view of brain anatomy as the basis for all mental phenomena. Unlike Édouard Séguin, he sought the causes of retardation, insanity, and even talent in the folds of brains. It was Séguin's wish that Spitzka make some use of his own brain, and it became the first of dozens of important brains that would come into Spitzka's possession.

Prior to the death of Séguin, the only American brain of note to have been removed and studied for a possible connection between anatomy and biography was that of Chauncey Wright, an eccentric mathematician, philosopher, failed Harvard lecturer, and founder of a short-lived dining society, the Metaphysical Club, whose other members included Oliver Wendell Holmes and William James. Wright died in 1875 at the age of forty-five, and his brain proved to be an embarrassing exception to prevailing ideas about fissural development and mental acuity.[3] It was big enough, at 1,516 grams, to satisfy believers in the relation of brain weight to intelligence. But it was even more remarkable for the simplicity of its fissures and its smooth exterior. Wright's was a confounding case—a man of undisputed intelligence, even genius, whose brain, if judged by appearance alone, would make him out to be an idiot, or at best an imbecile (in the technical sense of the terms). If nothing else, it suggested that more study was needed, and Spitzka decided to add this line of investigation to his many other pursuits.

Although he was a native New Yorker, Edward Spitzka had inherited his father's German mentality, which is to say, he harbored a barely concealed disdain for American institutions and scholarship. His own education had been impressive, not only in science but in languages, history, and literature. He was a Renaissance man in what he considered to be a backward nation, and he wasted no time in establishing a reputation.

Before he was out of his twenties, Spitzka was known nationally as an expert on insanity and as an accomplished anatomist. He would later gain some unwanted notoriety when, in his mid-thirties, he presided over the country's first execution by electricity. It fell to Spitzka to decide how long to apply the current. With no precedent to draw upon, he halted the procedure before the subject, a murderer named George Kemmler, was dead. A subsequent application of current turned Kemmler into a burnt and convulsed corpse. Horrified onlookers blamed Spitzka, who tried to distance himself from the fiasco. Even so, he was not too horrified to perform the autopsy and to remove Kemmler's brain.[4]

Well before the Guiteau trial, Spitzka had also made headlines as a social reformer and ardent critic of institutions for the insane. In 1878, he fired the opening salvo in a war that erupted between a group of New York neurologists and an association of psychiatrists over who should control the asylums and their burgeoning populations. For Spitzka, the stakes went beyond mere money and power, or even principle. He understood that whoever ran the asylums would have access to an almost inexhaustible supply of brains of the insane.

WHEN THOMAS WILLIS coined the term *neurologie* in 1664, he could not have known how far-reaching the word would become. During the nineteenth century, with the acceptance of Gall's idea of the brain as the organ of the mind, and with the rise of scientific materialism, *neurology*, a term designating the scientific study of the human nervous system, came to embrace all mental phenomena, especially diseases of the mind.

Edward Spitzka was one of a new breed of neurologists—men who considered themselves positivists, freethinkers, and materialists, and who were convinced that clinical observations should be supplemented by postmortem examinations of brains. Most mental diseases, he believed, had an organic basis, and were most likely caused by lesions of the nervous system. One of Spitzka's earliest research papers, from 1876, was entitled "The Somatic Etiology of Insanity" (loosely translated, it means "the physical causes of insanity"), and it won him a prestigious prize from the British Medico-Psychological Association.

In Spitzka's day, medical research in the United States was done by enterprising physicians in their spare time. Medical schools did not maintain laboratories, and placed greater value on clinical practice than on research. Experimentalists like Spitzka often found it difficult to land professorships, and they were often dismissed as mere "book-learners."[5] Unable to land institutional backing for his anatomical research, Spitzka supported himself through his private practice as an alienist, essentially as a physician specializing in nervous diseases, and he supplemented this income by lecturing at medical and veterinary colleges in New York, and by serving occasionally as an expert witness in trials involving criminal insanity. But his overriding interest was the organic basis of insanity, a specialty that had not yet been recognized by the medical schools, and in order to conduct this research, he needed access to "material," specifically to brains of the insane, most of whom lived and died at asylums.

One of the first state-run institutions for the insane opened in

Utica, New York, in 1843, under the benevolent directorship of Amariah Brigham, a physician who viewed insanity as a curable disease of the brain. (Brigham went so far as to send to Calcutta for cannabis, which he used experimentally on his patients.) Like the mentally retarded, the insane had previously been kept out of sight—either at home or in poorhouses and prisons—and effectively out of mind, until some enlightened physicians took an interest in their plight. At the dawn of the eighteenth century, Benjamin Rush, a signer of the Declaration of Independence and the first great medical man of Philadelphia, campaigned to get the insane out of the poorhouses. But the problem required institutional measures that would not come about until midcentury. Private asylums existed for those who could afford them, but they housed only a small percentage of those in need. Utica was one of the first, and easily the grandest, of about a dozen state institutions built in the 1840s to address the problem.

The creation of the asylums brought a new profession into existence, that of medical superintendent. Like Brigham, most of the asylum directors were physicians with no special training either in insanity or in business administration. They learned their trade on the job and in isolation from the medical profession. In 1844, they banded together as the Association of Medical Superintendents of American Institutions for the Insane (AMSAII), which survives today as the American Psychiatric Association. Brigham was elected as the first president, and he also founded the *American Journal of Insanity*, the society's in-house publication.

By creating their own association, the superintendents did not intend merely to build an effective network of communication. They also created a closed shop. By limiting membership to asylum administrators only, they effectively excluded neurologists, pathologists, and clinical psychologists from their ranks. Even staff psychiatrists were barred from their meetings. With almost no oversight by the state or by the profession of medicine, the system was ripe for abuse. When

Brigham died in 1849, the Utica asylum was taken over by John Perdue Gray, who would run it like a corrupt ward boss.

In 1878, Edward Spitzka launched his first attack on the psychiatrists (the word then applied specifically to asylum directors) in a series of articles published in the *Journal of Nervous and Mental Disease,* the publication of record for the recently formed New York Neurological Society. In addition to exposing a litany of abuses, he attacked the superintendents as political hacks who were appointed "on grounds of nepotism and political favor."[6] He singled out Gray as an "indifferent, superficial man, owing his position merely to political buffoonery."

Spitzka saw the asylums as a squandered opportunity. In responsible hands, they would have made an ideal laboratory for the study of the physiology of the mind. In Europe he had worked in such institutions. The Salpetrière in Paris, for example, and the Bicêtre asylum maintained close relations with the teaching hospitals. Scientists like Broca and Manouvrier, even Lombroso, could conduct clinical examinations of neurological cases, and follow up with examination of brains. Spitzka was especially interested in bringing the asylums into the larger prospect of the professionalization of American medicine. He recommended that medical students be required to serve residencies at the asylums, that the clinical and pathological study of insanity become part of the portfolio of the general practitioner. In the current system, he noted, the superintendents did not carry out autopsies at all (perhaps not wanting to reveal how many patients had died of abuse), nor did they conduct scientific research. Spitzka mockingly praised them for their proficiency at gardening, farming, tin roofing, and drain pipe laying. They were, he said, "experts at everything except the diagnosis, pathology, and treatment of insanity."

As for their treatment of patients, he was outraged. After inspecting several facilities, he grimly shared his findings with an increasingly appalled public.

> What of an asylum whose record of accidents during the current year, a record which is by no means exhaustive, sounds like the list of casualties of a Bulgarian campaign? Three patients beaten to death, one of whom had twelve ribs broken! One patient boiled to death, by having hot water turned on him in a bath, while the attendant went out of the asylum building, leaving the helpless paralytic to his horrible fate, and several patients drowned, by falling off the asylum dock, in epileptic convulsions![7]

In response, the superintendents attacked the messenger. They argued that insanity was not a disease of the brain, at least not in all cases, and that it should not be made the province of self-appointed experts. Spitzka, they said, was stirring up trouble in order to advance his own career.

What began as a professional feud quickly became ugly enough to interest the press. While deploring the neurologists' personal attacks on the superintendents, papers like the *New York Times* agreed with Spitzka that the asylums were indeed sordid places, and that many of the directors were corrupt and uncaring. Spitzka was supported in his crusade by Édouard Séguin's son, the neurologist Edward Constant Séguin, the first president of the New York Neurological Society, and a fellow agitator, William Hammond, a prominent and high-priced alienist who courted derision from the superintendents because of his exorbitant fees for private consultations. As the feud escalated, principally between Gray and Spitzka, the neurologists succeeded in getting the state to appoint a commission to look into the matter.

At a hearing held in 1879, Spitzka presented a list of proposals for reform, including mandatory autopsies on all patients who died at the asylums. He again trotted out a litany of abuses, citing a recent visit to an asylum where "patients had pistol-shots fired into them by the superintendent, where patients were hung up by their wrists for hours, where the wards were a stench to the nostrils, the food unwholesome, the bedding covered with vermin, and the closets reeking with filth . . . where attendants freely confessed that the superintendent had 'blackened the eyes' of patients, and that they did not think anything of following his chivalrous example."[8]

But in the end, the commission ruled in favor of the superintendents. As Hammond ruefully remarked, they "could not have been more favorable to the superintendents if the latter had themselves, as was very likely, selected the names of the appointees."[9]

The neurologists were hardly blameless in defeat. The abuses they exposed were appalling, but so too were their personal attacks. Then there was their own Achilles' heel: the practice of offering expert testimony at criminal trials. How, the superintendents asked, is the public supposed to believe in neurology as an exact science when two neurologists can come to opposite conclusions after examining a defendant? The argument caught both Spitzka and Hammond off guard, and they did not have an answer. After losing the battle before the commission, the neurologists then proceeded to lose the war when both Spitzka and Hammond took the stand at the trial of the century, that of the assassin Charles Guiteau.

GUITEAU'S TRIAL BEGAN on November 14, 1881. Spitzka made his appearance a few weeks later. As he took his seat in the witness box, he knew what to expect. The district attorney, George Corkhill, had assembled a team of four veteran prosecutors with decades of political experience among them. A special counsel, Walter Davidge, Washington's most prominent attorney, was the star of the group. Corkhill had also hired Utica superintendent John Perdue Gray as a star witness and consultant, and kept him nearby so that he could feed questions to the team as they grilled Spitzka. The wounds from the asylum commission hearing were still raw, and Gray was out for revenge.

On the other side of the aisle, Guiteau was represented by his brother-in-law, George Scoville, a midwestern lawyer who had never tried a murder case. He would call several experts to testify to Guiteau's insanity, but Spitzka was his only real hope.

Because he taught at a veterinary college (a fairly typical assignment for an anatomist), Spitzka expected his credentials to be challenged, and he was not disappointed. After patiently trying to explain the vagaries of academic research, he was asked point-blank by Davidge, "You are a veterinary surgeon, are you not?" To which he testily replied, "In the sense that I treat asses who ask me stupid questions, I am."[10] (The reply brought howls of laughter from the gallery.) To the charge that he was an atheist, he replied, "I refuse on principle to answer such a question; it seems to me impertinent in a country that guarantees civil and religious liberty." Accused of being an expert for hire, he pointed out that the prosecutor himself had tried to acquire his services, and that the fee he was getting was not enough "to take me to Jersey City." At each of these exchanges, the gallery applauded, laughed, and cheered, while the prosecutors fumed. They took turns trying to undermine Spitzka's credibility before giving up in frustration.

Spitzka's debt to Lombroso came out most clearly in testimony elicited by a skeptical Walter Davidge, whose questions were given to him by John Gray. Much of Spitzka's assessment of Guiteau was based on a physical examination and a brief interview. In the asymmetry of the killer's face, the shape of his head, the deviation of his tongue, his lopsided smile, Spitzka detected a probable deformation of the brain, and concluded that Guiteau suffered from a form of organic disease, a mental sickness that rendered him morally insane. "Those were the evidences that I found that he was born with a brain whose two sides are not equal," he testified. In his expert opinion, Guiteau was a "moral monstrosity," or, what was the same thing, a "brain monstrosity."

> By a moral monster, I mean someone who is born with so defective a nervous organization that he is altogether deprived of that

> moral sense which is an integral and essential constituent of the normal human mind, he being analogous in that respect to the congenital cripple who is born speechless, or with one leg shorter than the other, or with any other monstrous development that we now and again see.

Gray, testifying for the prosecution, called this nonsense. Guiteau had studied the law, he pointed out, and had passed the bar. His intellectual faculties were obviously intact. Gray refused to accept the phrenological claim, proposed years earlier by Samuel Gridley Howe, that the mind has both a moral faculty and an intellectual faculty, and that one of these can be extinguished without affecting the other. He insisted that the mental faculties are all one and the same, that disease cannot isolate and affect one faculty and not others. As far as he was concerned, moral insanity did not exist.

Spitzka did not base his testimony entirely on anatomy. He also took Guiteau's history into account by alluding to the fact that the defendant had recently tried to kill his own sister-in-law. But in the end, Spitzka's bravado and the scientific arguments didn't matter. The neurologists' position was doomed when it became allied to a man who had shot and killed a president.

Charles Guiteau was convicted and sentenced to death by hanging. Afterward, his body was examined and his brain removed by Daniel Smith Lamb, who had also performed Garfield's autopsy. Spitzka was prevented from attending the postmortem by the prosecutors (probably at Gray's insistence), but when he read the autopsy report, he felt vindicated.

THE BRAIN OF Charles Guiteau (most of it, anyway) now resides in the National Museum of Health and Medicine, which also has Abraham Lincoln's brain. It shows no evidence of disease, and gives little insight into the warped mind of an assassin. Even so, the autopsy report contained a few details that led some of Spitzka's doubters to change their minds about the case. Guiteau seems to have been a walking showcase of degenerative afflictions. These included pathological changes—"degeneration of gray cells and small blood vessels"—that could have been signs of insanity (they have no such significance today). The report also revealed that Guiteau suffered from syphilis, which he undoubtedly picked up from a prostitute (by his own admission), and his lungs showed signs of tuberculosis.[11] He was hardly a picture of health, another point that Spitzka was pleased to note.

From a modern perspective, there is nothing in the report that can explain Guiteau's behavior in brain-pathological terms. Nor did his brain vindicate the neurologists in their war with the psychiatrists. In the public's mind, the trial merely confirmed their suspicion that neurology was more art than science.

TODAY NEUROLOGY is a well-established science, even though it's founding premise has fallen by the wayside. It is suspected that some mental disorders do correlate with brain abnormalities. The brain of convicted mass murderer and cannibal Jeffrey Dahmer was examined for just such signs of defect.[12] But even where such defects are found, it is not clear what role they may play in turning thoughts into deeds. In 1913, the role of neurochemicals in behavior was first proposed, and it tilted the mind-brain discussion from gross anatomy to physiology, principally toward the role played by hormones and neurotransmitters. But this did not quell the fascination with brain structure, which continues to induce researchers to crack open skull caps when they have the chance.

Dahmer's case is not unique. The same fate befell members of the notorious Baader-Meinhof gang, Germany's terrorists of the 1970s, whose brains were taken after their deaths in prison, and whose study has not yet yielded any results.[13] It seems that whenever the opportunity arises to examine such brains, someone usually takes it, although the motivation usually has less to do with the clinical aspects of the case than with its newsworthiness. Once the story fades from the headlines, the brains are usually forgotten.

Such was the case with Charles Guiteau. No present-day neurologist would suggest that his pickled brain can shed any new light on the events of 1881. Today, its publicity value is almost nil. In the wake of the Guiteau trial, on the other hand, it was the talk of the town.

Two years after the trial, Spitzka published a textbook, *Insanity: Its Classification, Diagnosis and Treatment,* which exposed the weakness of the neurologists' position. At the outset, Spitzka had to concede that "in the present state of our knowledge, it is impossible to frame a definition of insanity which, while it meets the practical everyday requirements, is constructed on *scientific* principles." Nor could he and Hammond solve the problem of classifying patients by degree or type of insanity. They had hoped to provide such a classification based on morbid anatomy, on the correlation of brain lesions with behavioral disorders, but they ran up against the old phrenological conundrum. "It is to be regretted," conceded Spitzka, "that the present state of our cerebral anatomy and physiology is such as to prevent our making any precise localizations of the several forces and faculties which go to make up the mind."[14]

"Unfortunately," Spitzka continued, "sufficient positive observations on which to base a thorough pathology of the diseased states are, as yet, desiderata." In other words, they needed more brains, and in the coming decade, finding them would become Spitzka's personal crusade.

Chapter 10

Wilder

IN THE FALL of 1889, five distinguished scientists met at an elegant row house near Philadelphia's Society Hill for the purpose of launching an unusual enterprise. Their goal was to create a new scientific society, to add one more to the long list of clubs to which they already belonged. Privately, they admitted that these societies were little more than an excuse for indulging in good conversation, for downing bowls of turtle soup and goblets of Madeira. But this society would be different, if only in the degree of commitment it would entail. Following in the footsteps of the French materialist dining societies, the Philadelphia men had decided to bequeath their brains to each other.

The house belonged to William Pepper, the provost of the University of Pennsylvania. At the time, Pepper was one of the most celebrated physicians in the country. Five years earlier, at the age of thirty-nine, he had taken the helm of the university, intent on reshaping it into one of the finest centers of professional education in the world, and he had succeeded, particularly in the School of Medicine, whose curriculum had attained European standards. Three of his faculty members were on hand that evening to help Pepper launch the new society. They were Francis X. Dercum, a specialist in nervous diseases; Harrison Allen, a comparative anatomist; and Joseph Leidy, the president of the Academy of Natural Sciences, and perhaps the country's preeminent natural scientist. (A friend of Pepper's had once wryly noted, "Never give Leidy anything for dinner that he can dissect." At a dinner party he had served Leidy nine turtles. Instead of eating them, Leidy proceeded to dissect them, finding and naming three previously unrecorded intestinal parasites in the process.[1])

Pepper, Allen, Dercum, and Leidy were important men not just in Philadelphia, but also in American science. Following in the tradition established by Benjamin Franklin, they used scientific societies as clearinghouses for new discoveries. Through the prestigious American Philosophical Society (founded by Franklin in 1743), and the Academy of Natural Sciences (founded in 1812), they lured new talent to the city. The university in particular had become the center of medical education and research in America. Just two years earlier, in another of their speculative ventures, the four men had supervised Eadweard Muybridge's famous photographic studies of animal locomotion. Dercum provided handicapped patients and other subjects from Blockley Hospital, Allen helped to procure animal subjects, and Muybridge photographed them as they executed a range of motions. The sequential images he produced would, among other things, lead to a revolution in modern art, and serve as the precursor to motion pictures. Pepper funded the studies, Leidy served on the advisory board, and Allen and Dercum contributed monographs to accompany the final publication. They may not have contributed much to anatomical science, but Muybridge's motion studies were typical of a wide range of open-ended explorations going on in Philadelphia at the time. If it wasn't clear what they would lead to, they were certain to produce some interesting results. The same could be said for the new brain donation society.

The fifth and youngest member of the group that convened at Pepper's house was in many ways the odd man out, but his presence was essential, not only because he had started the first elite brain collection in America, but also because he was the country's top brain anatomist. This was, of course, Edward Charles Spitzka.

Although the meeting was not well documented, it seems that in forming an American mutual autopsy society, the Philadelphia men were not driven by a political agenda. Unlike their French counterparts, they were not fighting for academic freedom. Theirs would be a society of eminent men committed to studying the brains of other eminent men—in essence, their own brains. They decided to call it the American Anthropometric Society (AAS), partly to indicate the scope of their crusade, partly to advertise its affinity with physical anthropology, and mostly to give it the stamp of legitimacy. Reporters had trouble with the name from the start, and soon resorted to calling it the Brain Society. It wasn't long before they were referring to the membership, perhaps ironically, as the brain men.

THE REMNANTS OF the Brain Society, which is to say the brains of its members, can be found today in a basement closet at the Wistar Institute of Anatomy and Biology, a West Philadelphia landmark that has stood at the corner of 36th and Spruce Street, in the heart of the University of Pennsylvania campus, for over a century. Hemmed into a wedge where Woodlawn Avenue cuts a diagonal swath through the grid of the city's streets, the place is an awkward amalgam of old and new, the kind of place that inspires more dread than curiosity. The old part consists of a cold, beige brick, and somewhat eerie Victorian pile; the newer half a cold, red brick, and eerie mid-sixties annex.

The Wistar Institute was built in 1894 to house the Wistar-Horner Museum, the anatomical collection of the university's medical school at the time of the Brain Society's founding, and it remained, essentially, a state-of-the-art mausoleum until the 1980s, by which time the institute had long since branched out into polio vaccines and cancer research. And yet for most of the twentieth century the bulk of its space was dedicated to the display of human and animal body parts, many of them bearing witness to the imperfect mechanism of genetic transmission, the opportunistic cruelty of bacteria and viruses, and the horrors of cell mutation. Here were once exhibited, for the edification of a gaping public, pickled babies with cleft lips and club feet, Siamese fetuses, venereal deformations, tubercular organs, and the ravages of plagues and tumors. But the brains of eminent men were never displayed for the simple reason that, with two notable exceptions they are the only specimens in the entire collection that can be identified with specific people. The articulated skeletons, for example, have no names. The Siamese fetuses never had names. The brains, on the other hand, all twenty-two of them, may have succumbed to obscurity, but not to complete anonymity.

ISAAC JONES WISTAR, or more properly, Brigadier General Isaac Wistar, was the man who put up the money for the building and gave it his name. Wistar's spirit may no longer haunt the cramped corridors of the institute, but his body is still there—its parts dispersed throughout the building. In the second-floor lobby, in six display cases recessed into the walls, are urns containing the ashes of illustrious men from the institute's past, with a place of prominence set aside for the general. While leading the Seventy-first Pennsylvania Regiment at the Battle of Ball's Bluff in 1862, Wistar was shot in the right arm and his elbow joint fused (a condition known as anchylosis). Wounded again at Antietam, he was eventually forced to retire from active service.[2] The skeletonized right arm, fingernails included, now resides in the basement. So does Wistar's brain. His ashes, on the second floor, have been added to a pilgrimage route of graves of Civil War commanders.

The Wistars were and are one of the first families of Philadelphia society. Isaac's great-uncle, Caspar Wistar, was the third president of the prestigious American Philosophical Society (after Benjamin Franklin and Thomas Jefferson). A contemporary of Benjamin Rush, and like Rush a hero of the yellow fever epidemic of 1793, Wistar is today memorialized (more than remembered) through the wisteria vine (originally the *wistaria*), named for him by his good friend the Philadelphia naturalist Thomas Nuttall. His other legacy is the Wistar Anatomy Collection, which he began while teaching at the University of Pennsylvania in the early 1800s. Many of Wistar's original anatomical preparations are still in the collection. In fact, Wistar's own heart is there in a cardboard box. It is the premodern, which is to say pre-phrenological, analog of the brains nearby—the seat of Caspar Wistar's soul, hardened to the consistency of a dog's chew toy and wrapped in tissue paper.

Isaac Wistar did not follow in his great-uncle's footsteps. Neither a physician nor a scholar, he made his fortune after the war in mining and railroads. The institute became his hobby, and while serving as chairman of the board of trustees for fifteen years, he tried to steer its mission toward brain research. His own brain testifies to his commitment to that plan. When he died in Delaware in 1905 at the age of seventy-eight, his doctor, Alfred Stengel, and the institute's director, Milton Greenman, rushed out to the family farm, laid the general out on the kitchen table, cut off his right arm above the elbow, sawed off his skull cap, removed his brain, sewed him up, and rushed back to the institute with their prized specimens. In time, the brains of Greenman and Stengel would also make their way to the basement storage room.

The urn next to Isaac Wistar's contains the ashes of Joseph Leidy. His brain, also in the basement, is considered the prize of the lot. When he died in 1891, Leidy was the most famous scientist in America, renowned for the breadth as well as the depth of his knowledge. Not a theorist, but rather the last great generalist in the tradition of Agassiz and Cuvier, he assumed the chair of anatomy at Penn when he dissected his predecessor, William Horner (after Horner's natural death, of course), and took over the Wistar-Horner Museum. Over a long career, Leidy became a walking encyclopedia of scientific knowledge. It was said that he could turn over any rock and name every creature under it, as well as the rock itself and the mineral composition of the soil.[3]

Edward Drinker Cope is there, too. Cope is best remembered today for his rivalry with Yale's Othniel Marsh. In their attempts to pillage the Montana and Wyoming dinosaur graveyards in the 1890s, the two men engaged in one of the most notorious scientific feuds of all time. It got so bad at one point that Leidy, a better scientist than either and a mentor to both, quit paleontology in disgust at their bickering and underhandedness. Cope's life was marked by a lack of decorum. "No woman was safe within miles of him," a friend once noted. "His mouth was the most animal I've ever heard," said another.[4] Yet he was also a highly respected scientist who authored over twelve hundred publications, in which he identified and named many of the great extinct reptiles. The infamous bone hunter cheated the undertakers by bequeathing both his brain and, appropriately enough, his skeleton to the Wistar Institute. His dinosaur bone collection is now housed at the natural history museums in New York and Washington, D.C. His own skull and bones occupy a cardboard box kept out of sight at the university's archaeological muscum. His ashes, or at least the ashes of what little was left of him, are in an urn near those of Wistar and Leidy. His brain keeps company with theirs in the basement.

The other brain men represented in the Wistar collection share a dubious distinction. They lived and worked at one degree of separation from everlasting renown, and it earned them, as one of them put it, the "immortality of a decade." To give them thcir duc, most of them had lasting legacies. The membership roll reads like a gallery of forgotten patriarchs: Cope, the father of ichthyology and herpetology; G. Stanley Hall, the father of American psychology; William Osler, the father of modern clinical medicine and inventor of bedside manner; Joseph Leidy, the greatest American scientist of the mid-nineteenth century and, in the words of his biographer, the "last man who knew everything";[5] William Pepper, the steward of the University of Pennsylvania and the Philadelphia Free Library; J. B. Haldane, father of the geneticist and essayist J. B. S. Haldane, and a pioneer in the study of human respiration; and Henry Herbert Donaldson, the man who studied the brain of Laura Bridgman, and who introduced the albino Norway rat as the laboratory animal of choice throughout the world.

Not surprisingly, several brains in the collection have gone missing. Those of Édouard Séguin's colleague Isaac Newton Kerlin and his wife, Harriet, for example, belong in the Wistar collection, but seem to have been lost en route from the Elwyn Institute. Another specimen that has haunted the Institute (if only in a figurative sense) from the very start is the brain of Walt Whitman, which was removed, preserved, and sent off to the Wistar Museum upon his death in 1892. It was later said to have been destroyed when a careless laboratory assistant dropped the jar containing it. The story is a dubious one (as will be shown), but it has a delicious irony. Although Whitman was an outsider with no academic credentials, his was the one brain that might have put the collection on the map.

THE WISTAR COLLECTION is not the only repository of eminent brains in America. Some two hundred miles north of Philadelphia, on the campus of Cornell University, a similar collection bears witness to a schism within the American Anthropometric Society that occurred in 1891, and it highlights an awkward fact about elite brain collections—their ideological value always seems to trump their scientific value. The Cornell collection was not created for the purpose of finding the anatomy of eminence. Instead it was dedicated to the premise of studying brains of "educated and orderly persons." It may sound like a mere quibble over semantics, but the distinction would end up costing the AAS hundreds of brains.

At one time, the Cornell collection contained 430 brains of adults and children, and over two hundred brains of embryos and fetuses. Only twenty-six remain, most of them crammed (no surprise here) into a storage closet in the basement of Uris Hall, the home of the psychology department. There are eight notable exceptions, currently on display in a brightly lit cabinet in an otherwise drab corridor on the second floor. They include the brains of Edward Bradford Titchener, a pioneer in American psychology; of Daniel Smith Lamb, the man who performed the autopsy on President Garfield and removed the brain of the assassin Charles Guiteau; and of Edward Howard Ruloff, who had the distinction of being a man of letters and of mayhem. Ruloff's is by far the biggest, and easily the most interesting, specimen in the exhibit.[6]

Edward Ruloff was a thief and a murderer, a career criminal, perhaps even a criminal genius. He was also a respected philologist who published a well-received article entitled "Method of Formation of Language," while he carried out a series of petty crimes in upstate New York. During a crime spree that spanned his entire adult life, Ruloff managed to prevail in a trial for the murder of his wife and son (a crime he most likely committed), and was later convicted of robbery. He escaped from jail, even managed to elude a lynch mob, was enthusiastically recommended for a teaching post at Duke University, but then decided to stay in New York. After another heist, two of his accomplices washed up in the Chenango River, and Ruloff was arrested again. He stood trial, not for the murders of his colleagues, which he probably committed, but for killing a store clerk during the robbery. He insisted on conducting his own defense, but his luck had finally run out. He was executed by hanging in 1871. His brain, which weighed 1,770 grams, is not only the largest brain in the cabinet, but one of the heaviest preserved brains extant.

Officially, the eight brains on display at Cornell (including, somewhat playfully, the brain of a pumpkin) are referred to as the Wilder Brain Collection, after the man who started it all. Burt Wilder, an anatomist of great distinction, was a charter member of the American Anthropometric Society, and one of the club's most enthusiastic participants. He was also the first, if not the only, member to resign in protest.

BURT GREEN WILDER was a very different sort of brain expert than his Philadelphia counterparts at the Brain Society. A pious man, a tireless crusader for social justice, and a man of artistic sensibility, Wilder was a missionary on behalf of proper hygiene and conduct. Each fall, he gave a mandatory lecture at Cornell entitled "What Every Young Man Should Know." He actively opposed smoking, gambling, and intercollegiate athletics. He set Joyce Kilmer's and Oliver Wendell Holmes's poetry to music. He harangued the novelist Owen Wister for making denigrating (and erroneous) statements about African Americans in his novels, and berated William Dean Howells for including gratuitous smoking scenes in his fiction. From 1863 to 1865, Wilder served in the Union Army as a surgeon in the Fifty-fifth Massachusetts Infantry, a black regiment formed after the famous Fifty-fourth, which was led by Wilder's good friend Robert Gould Shaw. After his service with the Fifty-fifth, Wilder became an outspoken advocate for civil rights, a rarity among men of science. He never missed an opportunity to express his deep admiration for the men of his regiment, caring little whom he might offend. When rebuked, he was happy to explain himself, but he never apologized. Burt Wilder knew he had the truth on his side.

Wilder arrived at Cornell in September of 1867 after finishing a degree in comparative anatomy at Harvard, where he served as an assistant to the great Swiss naturalist Louis Agassiz. At the age of twenty-six, he was appointed Cornell's first professor of anatomy, and he was faced with the monumental task of building a department and a laboratory entirely from scratch. Although he was interested in the study of the human brain, specimens were scarce and expensive. In 1886, he persuaded the college to purchase the brain, hands, and face of Edward Ruloff for fifteen dollars. He also obtained, for the sum of fifty dollars, the carcass of a baby gorilla, the first live gorilla ever brought to North America. (It died in Boston five days after its arrival. Wilder preserved the brain, skeleton, and the skin, which he had stuffed. They have since disappeared from the Cornell collection.) But these were exceptional acquisitions, and not the everyday material of a teaching laboratory.

Given few resources with which to acquire specimens, Wilder made a virtue of necessity by taking the advice of Joseph Leidy, who once said to Wilder's good friend Harrison Allen: "The anatomy of the domestic cat closely resembles that of a man. If you cannot dissect your neighbor's body, steal and dissect his cat."[7]

Rather than steal cats, Wilder put out an appeal to the public. He posted flyers around Ithaca offering twenty-five cents to any boy who would bring him an animal. The plan was a fantastic success and soon rid Ithaca of its stray cats, and many of its domesticated ones. According to one reporter, Ithacans openly thanked Dr. Wilder for allowing them "a comfortable night's sleep when the citizens of other towns are leaning out of their windows seeking a chance to kill the members of some cat congress."[8] To accommodate the influx of animals, Wilder constructed a cat house (in the literal sense of the term) near the anatomical laboratory, and placed a sign over the door to the pens reading:

SNUG HOUSED AND FULLY FED,

HAPPY LIVING AND USEFUL DEAD.[9]

The casual passerby might well have wondered, useful for what?

According to the *Ithaca Journal*, Wilder was the first professor in America to practice vivisection in the classroom. Simply put: "To illustrate the action of the heart and lungs he has put hundreds of cats under the influence of chloroform and cut them up in his lecture room."[10] The practice would lead to a massive protest movement among the very people who supported Wilder's progressive social ideas. It would also lead to Wilder's first major work, a well-received study entitled *The Brain of the Cat.* But this was merely a prelude to a more ambitious agenda. Having exhausted the possibilities of cats (not to mention the local cat population), Wilder got the idea that he might induce the great minds of the university to hand over their brains after they were done with them.

The plot was hatched one night in 1889 over dinner with friends at his lodging house. None of the participants remembered exactly who suggested it, but one of them, a history professor named Goldwyn Smith, went so far as to draw up a bequest then and there. Being possessed of an evangelical frame of mind, Wilder modeled his Cornell Brain Association on the temperance movement rather than the blood-oath societies that were popular at the time. He drafted a rousing speech and an impressive bequest form, which read, in part:

> I, ______, recognizing the need of studying the brains of educated and orderly persons rather than those of the ignorant, criminal or insane, in order to determine their weight, form and fissural pattern, the correlations with bodily and mental powers of various kinds and degrees, and the influences of sex, age and inheritance, hereby declare my wish that, at my death, my brain should be entrusted to the Cornell Brain Association.[11]

At social functions attended by Cornell faculty and alumni, Wilder would stand and deliver his plea, then hand out the bequest forms. To anyone who took the pledge, he promised to send "provisional diagrams of the fissures," as well as copies of any publications that resulted from other donations. Many took him up on the offer, including the suffragist Elizabeth Cady Stanton (whose brain was not collected) and the neuroanatomist C. Judson Herrick (whose brain would be the last to enter the collection, in 1960).

At the outset, Wilder and his friends did not think to form a society. They were content to set in motion a plan for acquiring brain specimens. So when the American Anthropometric Society was founded later the same year, Wilder saw no conflict of interest and joined at the first opportunity. He was promptly appointed head of the publications committee.

Wilder correctly assumed that the Philadelphia brain men were motivated by the principle of cerebral localization, by the desire to create a new phrenology and to place it on a scientific footing. Their plan sounded simple enough: they would assemble an archive of exemplary brains that would someday provide the data to complete the job that Gall had tried to accomplish with busts and skulls. They would not make the mistakes of their European predecessors. They would not focus on genius, nor would they try to define intelligence. But as Wilder soon learned, it was not as though the brain men had no agenda at all. If their plan succeeded, and they assumed that it would, they hoped to confirm the anatomical superiority of eminent men like themselves.

Wilder had two problems with this agenda. The first and most important was proprietorship. The AAS proposed to act as a clearinghouse for brains that had been languishing in private hands. Acquiring Wilder's budding collection (at that time the country's best), along with his pile of brain bequests, would have been a coup. But the Cornell brains had to stay at Cornell as a condition of the bequests. "No other institution or local entity," Wilder wrote to his fellows, "is entitled to absorb or appropriate in any sense the encephalic material or reputation, present or future, of this University."[12] Wilder's other point of contention involved classification.

The Philadelphians professed their interest in "high-type brains," as they liked to call them. Officially, they spoke of "brains of eminent men," by which they meant white, professional men of good social standing. Wilder rejected this narrow definition. Among the brains he had acquired at Cornell, Edward Ruloff's was by far the biggest, and clearly it had belonged to an eminent man of sorts, although Wilder would not have called him a "high type." As he saw it, there were equally fine minds among members of other races, among the opposite sex, and within all social classes. But these were given short shrift by the Philadelphia brain men. Wilder was especially concerned with the dignity of African Americans, in particular the soldiers of the Fifty-fifth Massachusetts Infantry. They were, in his estimation, the finest men he had ever known.

Wilder shared his concerns with Harrison Allen, who had assumed the presidency of the AAS after the death of Joseph Leidy in 1891. Unfortunately, Wilder's half of the correspondence is missing, but Allen's letters show that he shared many of the same misgivings. When Wilder submitted his resignation in December of 1891, Allen graciously replied, "I hope that you will remain as a friend and counselor, for science knows neither Phila., Cornell, or any other place."[13]

Wilder then went forward with his own plan, which entailed two collections, one consisting of brains of "educated and orderly persons," as he labeled them; the other a collection of brains of criminals and the insane. Into the former cabinet went the brains of professors, as well as those of orderly men and women, both black and white. Edward Ruloff's brain went into the latter.

BOTH THE CORNELL and Philadelphia brain societies grew rapidly, and by century's end boasted scores of members. The brains themselves trickled in much more slowly. However enthusiastic the donors might have been at the time of their initiation, their surviving relatives often proved to be a harder sell. Few of the donors let their families in on their secret. There was also the delicate matter of timing. Death affords a twenty-four-hour window of opportunity at best, the brain being the first part of the corpse to decompose, so the decision to redeem the pledge had to be made quickly and executed swiftly.

The brain societies were an odd sort of undertaking (so to speak) in other respects, perhaps the most obvious being that acquisitions were hardly cause for celebration. When Joseph Leidy died in 1891, the American Anthropometric Society lost its first president, if not its most famous member. When Andrew J. Parker, a professor of biology at Penn, died a year later, the society lost its best young anatomist, and one of the few members who had the time, interest, and expertise to make a study of the brains. When Walt Whitman died just a week after Parker, a different problem surfaced. He was the first nonscientist to die, the first brain donor whose membership was contested by his family, the first controversial case for a society that never anticipated any controversy. Because of Whitman, the brain men would, for the first time, be faced with intense scrutiny and negative publicity, which forced them to become a much more secretive society.

Meanwhile Wilder's Cornell Brain Association remained an open book, its founder continued with his relentless proselytizing, and its members kept to their pledges. There were a few exceptions, but Wilder soon had more material than he could use, which led to an unanticipated problem. For all of the energy they expended in acquiring their precious specimens, the brain men had not given much thought to what they would do with them. Like his eminent colleagues in Philadelphia, Wilder had his hands in too many pies. He had no time to mount a thorough investigation of the brains, and so they sat on the shelves gathering dust.

BURT WILDER COULD easily be written off as a crackpot had he not been one of the foremost brain anatomists in the world. Although he worked in relative isolation, he quickly developed an international reputation. In 1892, he wrote an extensive entry on the brain for the

Reference Handbook of the Medical Sciences, the authoritative text of the day. As president of both the Association of American Anatomists and the American Neurological Association in the 1880s, he waged a spirited battle with German anatomists to reform the arcane Latin terminology used to name the parts of the nervous system. Outside of the laboratory, Wilder's enthusiasm could not sustain some of his grander ideas. He frequently found himself tilting at windmills, and nowhere was he more frustrated than in his crusade for racial justice.

In 1905, Wilder took offense at a scene in a new novel by the Philadelphia lawyer Owen Wister (a cousin of Isaac Wistar's), who is best known as the author of *The Virginian.* The novel—*Lady Baltimore*—was serialized by the *Saturday Evening Post,* and in one installment the main character compares the skulls of a white man, a black man, and a gorilla, and finds in the latter pair "a kinship which stares you in the face." Wilder at first assumed that Wister unknowingly recycled a myth that had been circulated years earlier by craniologists. He dashed off a polite letter of clarification, pointing out the error from the point of view of comparative anatomy. Wister would not have written such a thing, Wilder wrote, had he ever seen a set of such skulls.[14]

Wister responded rather dismissively through his secretary that the scene merely reflected common knowledge. An infuriated Wilder then dropped all courtesy and suggested that Wister should visit a museum and, after educating himself, should write a public retraction. After a few more rounds, Wister grudgingly changed the wording, but not the original sense of the passage, and the contest ended in a draw. It was Wilder's first foray into the moral and scientific education of the reading public.

Wilder continued to make civil rights one of his most important missions.[15] In 1909, while attending a conference of the National Negro Committee, he delivered a long paper, partly technical, but mostly sentimental, entitled "The Brain of the American Negro," in which he called for an end to discrimination based on supposed differences in brain structure. "Do any physical characteristics of the brain of the American Negro warrant discrimination against him, as such?" he asked. His answer, drawn from his experiences in the army and at Cornell, was blunt and courageous. "Respecting the brain of American Negroes," he wrote, "there are known to me no facts, deductions, or arguments that, in my opinion, justify withholding from men of African descent, as such, any civil or political rights or any educational or industrial opportunities that are enjoyed by whites." He was moved to add, "There have been many occasions when I was tempted to exclaim, 'Yes, a white man is as worthy as a colored man—provided he behaves himself as well.' "[16]

The only problem was that Wilder had not conducted any comparative studies using his collection. His paper on the brain of the Negro was not, in fact, a scientific study at all, and it relied on just a few cases. For his anatomical evidence, he relied on the work of other anatomists.

Throughout his life, Wilder would find virtue to be a hard sell. In 1915, a former student of his, Robert W. Shufeldt, published *America's Greatest Problem: The Negro,* one of the most offensive race-baiting books of a notoriously uncivil era. Wilder wrote a long and scathing review for the journal *Science,* at that time edited by the psychologist James McKeen Cattell. Cattell found the review too caustic to print in its entirety, and cut it to two blandly critical paragraphs.[17] He wrote Wilder to say that he did not think it proper to "take up in *Science* the question of the social relations, etc. of the Negro." The "etc." was a veiled reference to Shufeldt's belief in the predatory sexuality of African Americans.

SOME TWENTY YEARS after the founding of the two brain societies, Edward Spitzka was asked who had been the first to start a brain-donation society in America. He replied:

> At the time I thought that Pepper simply stole Wilder's thunder. Wilder undoubtedly was first in this country and I recall telling him of the Paris Society and that the fact was new to him. Old Séguin had spoken privately of such a thing, and his son had me get the brain in accordance with what he knew his father's sentiments were. And here's another curious circumstance: While at the autopsy, Amidon [Spitzka's assistant], speaking of old S's views, said: "I'm willing to start in as a member of such a body right now, if others will." These are plain facts; but Wilder is undoubtedly way ahead of Pepper, who merely took up mechanically what he'd heard.[18]

Like all of his predecessors in the field, Wilder hoped that the brains would yield immediate and obvious clues, and reveal a clear pattern. But they did not. He soon found, as had Wagner and Broca, that there were no telltale signs of high or low intellect, of criminality or morality to be found at first or even second glance. Faced with the seeming randomness in the folds of his pickled brains, Wilder realized that he would have to look deeper, which meant that the brains would have to wait.

Upon his retirement from Cornell in 1910, Wilder returned to his hometown of Brookline, Massachusetts, in order to take up the investigation he had put off for so long. He published a few reports, in which he limited himself to descriptions and refrained from speculations. When he died in 1925, at the age of eighty-three, his successor as curator of the Cornell Brain Association, James Papez, examined Wilder's brain and wrote a lengthy article in which the man and the brain elicit equal measures of praise.

"The Brain of Burt Green Wilder" appeared in the April 1929 issue of *The Journal of Comparative Neurology*. In it, Papez compares Wilder's brain, region by region, to those of twenty men and twenty women in the Cornell collection. The most detailed study ever made of the Cornell collection, it includes charts listing the lengths and depths of sulci, and areas of cortical regions of two dozen brains. Papez concluded that, consistent with other studies he had seen, the brains of scholars tend to have well developed frontal and parietal lobes, as measured by the depth and extent of the folds. Unfortunately, as even Papez had to admit, his observations were limited to the Cornell collection. He did not have a true control group.

Papez's study of Wilder's brain is typical of the genre—long on description, short on interpretation, and weakened by a dearth of confirming data. Papez would go on to do groundbreaking work in neurophysiology, but he would not attempt to complete Wilder's work on the hundreds of other brains at his disposal. Nor would anyone else.

WILDER AND THE brain men had one driving interest in common—they were consumed by the promise of phrenology, specifically by Comte's positivist vision of phrenology as a form of physiological psychology. But the phrenology they saw taking hold in America was very different from its European predecessors. They were competing with so-called "practical" phrenologists—bump readers who peddled a self-help philosophy with no real science behind it. Wilder fumed at the purveyors of this claptrap. As he saw it, these impostors had encroached on the anatomists' turf, and in yet another of his futile crusades, he challenged them in print, going so far as to have his chart of bumps read twice in an effort to debunk their claims. But his opponents easily outmaneuvered him by taking his criticisms out of context, and using them as endorsements of their product.

The product, in this instance, was the literary output of the Fowler brothers, Orson and Lorenzo, and their partner Samuel Wells, who were happy to provide answers where the brain men had none. The phrenological Fowlers were not men of science, but rather men of religion who applied a veneer of scientific legitimacy to a healthy-mind, healthy-body philosophy. When they burst onto the American scene in the mid-1800s, they filled a niche that was created when anatomists and physiologists failed to decipher the folds of the hemispheres. But unlike Gall, Wilder, or the Philadelphia brain men, the Fowlers cared little about the workings of the human brain, and they delivered on a different promise. Instead of trying to parse convolutions and fissures, they ran their hands over the crania of their customers, looked deep into their eyes, and in return for a modest fee, offered a glimpse into their souls.

Chapter 11

Whitman

PHRENOLOGY ARRIVED IN the United States in the 1820s, and gained a firm foothold in American culture when Caspar Spurzheim and George Combe came to New York in 1832 and set off on wildly popular lecture tours.[1] Spurzheim died within a few months, after having achieved national celebrity, while Combe spread the gospel of his Scottish brand of self-improvement through the same countryside that Alexis de Tocqueville had visited two years earlier while researching *Democracy in America.* But the event that sparked a phrenological industry and led to a distinctive, Americanized form of phrenology, had nothing to do with Spurzheim or Combe. It began instead with a school assignment.

In the spring of 1833, a young man named Henry Ward Beecher, a student at Amherst College, was assigned the task of arguing the anti-phrenological position in a college debate. Beecher hailed from an illustrious family. His father, the Reverend Lyman Beecher, was celebrated as "the father of more brains than any other man in America." Most of his twelve children rose to eminence, if not fame, and none more so than Henry, who would become the best-known minister in the country, and his sister Harriet Beecher Stowe, who wrote *Uncle Tom's Cabin.*

Beecher prepared for his debate by sending off to Boston for some literature from the American Phrenological Society, and he quickly became an expert on the subject. Although he easily defeated his opponent (he was, after all, destined to become one of the greatest public speakers of his generation), he became a convert to phrenology in the process. The tenets of self-improvement backed by hard science came as a revelation to him, and he sold the idea to his friend and fellow divinity student Orson Squire Fowler. Together, the two men began to offer cranial readings to their classmates at two cents a head.

With the arrival of summer, Beecher and Fowler expanded the operation, giving over two thousand readings in the surrounding towns and pocketing a handsome sum. But where Fowler saw opportunity, his partner only saw temptation. Faced with a crisis of conscience, Beecher decided to take the more righteous path offered by the Congregational ministry, and he enrolled in Cincinnati's Lane Theological Seminary, where his father served as president. Fowler wasted no time in recruiting his brother Lorenzo to take Beecher's place, and after graduation the two men set off on a yearlong lecture tour of New England, New York, Pennsylvania, and, ultimately, Washington, D.C. Wherever they went, they introduced the science of phrenology packaged as social reform and personal betterment, and word of their miraculous head readings spread like an epidemic.

In 1835, Orson and Lorenzo Fowler arrived in New York City and set up an office at 131 Nassau Street in lower Manhattan, which became the epicenter of a national movement. It began as a lecture agency, but the brothers quickly diversified and repackaged their product. They opened a bookstore and a "phrenological cabinet," and to fill its shelves they acquired skulls, elaborate charts, and the porcelain heads that would soon become a fixture in phrenological offices across the country. They also founded a publishing house, and to fill out its catalogue they began to translate the works of Gall and Spurzheim, to publish American editions of the works of George Combe, and, more important, to write phrenological tracts based on their immensely successful lectures. Among their first converts was a young medical student named Samuel R. Wells, who joined the firm, then the family (he married the Fowlers' sister Charlotte), and then the masthead. Thus was born the firm of Fowler and Wells, with offices in New York, Boston, and Philadelphia. It was only the beginning.

Initially, phrenology appealed to medical men more than the general public. Before the professionalization of medicine (which did not come along until the 1880s), American physicians dabbled in speculative remedies from patent medicines to bloodletting. According to some assessments, prior to 1850 a visit to a doctor was more likely to do harm than good. With its patina of scientific legitimacy, phrenology promised to bring order to this chaos. The first phrenological societies in America were in fact founded by physicians in the 1820s. During the next thirty years, while the movement was dying out in Europe, it took a firm hold on the American psyche. Everyone, in their way, became a phrenologist.

"The gross lines are legible to the dull; the cabman is phrenologist so far," wrote Ralph Waldo Emerson in *The Conduct of Life*. "A dome of brow denotes one thing, a pot-belly another. . . . People seem sheathed in their tough organization. Ask Spurzheim, ask the doctors, ask Quetelet, if temperaments decide nothing? or if there be anything they do not decide?"[2]

The list of influential Americans who had their phrenological charts read is a long one, and contains a veritable who's who of science, business, politics, and letters in the nineteenth century, including P. T. Barnum, John Brown, Louis Agassiz, and Samuel Morse.[3] But the name most frequently cited is that of the good gray poet, Walt Whitman, who walked a fine line between the practical phrenology of the Fowler brothers, which was more art than science, and the theoretical phrenology of American medical men. Whitman's is an interesting case because he wavered between the phrenological and anti-phrenological positions for most of his adult life. Unable to choose between the two sides competing for his soul, he was content to let the experts decipher not just the bumps on his skull, but the folds of his brain.

WALT WHITMAN'S FIRST encounter with phrenology occurred in 1846, when he reviewed Caspar Spurzheim's *Phrenology, or the Doctrine of Mental Phenomena* for the *Brooklyn Eagle.* He was already something of a medical gadfly. While growing up in Brooklyn, Whitman befriended many doctors and often visited surgical theaters at local hospitals. Anatomy and physiology fascinated him. Phrenology, on the other hand, seemed too good to be true. Here was a science that seemed to hold out a world of possibilities for social betterment, health, well-being, soundness of mind, and the celebration of the body. Yet it also had all the earmarks of bunkum.

Whitman ridiculed phrenology at first, but soon he tempered his criticism, conceding that "there can be no harm, but probably much good" in it. If he had his doubts about the product, he expressed a grudging admiration for the salesmen. "Among the most persevering workers in this country," he wrote, "must be reckoned the two Fowlers and Mr. Wells."[4]

Phrenology appealed to Whitman for the same reason it appealed to Americans in general. It confirmed his suspicion that the body was not something to hide beneath layers of gingham, flannel, and calico, but a gift worth celebrating. According to the Fowlers, sex was good. They had no hesitation in discussing it openly. Neither did Whitman, and his unrestrained verses scandalized his readers.

In the summer of 1849, Whitman decided to satisfy his curiosity. He boarded the Brooklyn ferry, walked straight to 131 Nassau Street, and was instantly bowled over. He had never seen anything like the extraordinary pastiche of skulls, books, curios, bones, pamphlets, and other oddities that awaited him at the Fowler and Wells Phrenological Cabinet. After several more visits, he finally screwed up his resolve to have his own chart read.[5]

During these visits, Lorenzo Niles Fowler had ample time to size up his man. He knew that Whitman wrote poems, although he did not know that he would soon become the nation's poet, the voice of an American renaissance. Fowler was in the business of assessing egos, not inflating them. The poets Oliver Wendell Holmes and John Greenleaf Whittier had come away from their own phrenological readings feeling less than flattered. But Fowler saw signs of greatness in Whitman. When the unknown poet walked through the door of his shop, he was an anybody; when he walked out, he was transformed. Lorenzo Fowler told him, in no uncertain terms, that he had the head and physique of a somebody.

In the hands of the Fowlers, a phrenological reading consisted of the palpation of thirty-five areas on each side of the head, and an assessment of six body characteristics. Each item was rated on a scale from 1 to 7, with 1 indicating a serious shortfall and 7 a possibly dangerous excess. Six was the ideal score. Lorenzo Fowler was impressed by Whitman on almost all counts. Out of the forty-one categories, he rated him in the 6 range on thirty-three of them.

It is an extraordinary chart, and Whitman instantly seized upon it as a validation of his uniqueness.[6] He would carry it with him for the rest of his life, reprint it a half dozen times in reviews of his own work and in prefaces to editions of *Leaves of Grass*, and from its distinctive terminology he would borrow such words as "amativeness," "adhesiveness," and "combativeness," and work them into his swaggering poetry. Whitman used his chart of bumps to celebrate himself: "Voluptuous, inhabitive, combative, conscientious, alimentive, of copious friendship," he wrote in "Song of the Broad Axe." In another poem he boasted: "My brain, it shall be your occult convolutions."

Whitman published the first edition of *Leaves of Grass* in 1855 at his own expense, and the firm of Fowler and Wells agreed to sell it through their bookstores and by mail order. There were not many takers. Whitman gave away more copies than he sold. He made a point of sending a copy to Ralph Waldo Emerson, not expecting but rather hoping for a blurb. He got one. In a famous letter, a prophetic letter, Emerson wrote, "I greet you at the beginning of a great career." He went on to call *Leaves of Grass* "the most extraordinary piece of wit and wisdom that America has yet contributed."

With a knack for self-promotion that could border on the obnoxious, Whitman ran the letter as a preface to the second edition without Emerson's permission. Emerson tried to distance himself, but too late. Fowler and Wells would also pull their backing, but Whitman simply brought out the second edition on his own. He never doubted his phrenological diagnosis, and, ever so slowly, his poetry began to sell.

American phrenology promoted the importance of the body and of the blood (which is to say, of heredity). It inspired ideas that would eventually lead, on the positive side, to vigorous therapeutic measures such as water cures, temperance, healthy diet, and an end to tight lacing; on the negative side, to eugenics. But its strongest selling point, and the innovation that would distinguish it from the preachings of Spurzheim and Combe, was its optimism. According to the Fowlers, the contours of the brain did not dictate an individual's fate. On the form they used as a diagnostic checklist for each of the phrenological faculties, they inserted column headings marked "cultivate" and "restrain" (as well as one headed "marry one having"). Which is to say that not only *could* every person work on their weaknesses by exercising their underdeveloped faculties and reining in their overdeveloped ones, but they had a moral obligation to do so. For a fee the firm of Fowler and Wells would tell them how to go about it.

Self-improvement was the crucial feature that distinguished the Fowlers' "practical" phrenology from Spurzheim's "theoretical" phrenology, and made it uniquely American. Instead of conceding the primacy of nature over nurture, and accepting the verdict of a phrenological diagnosis, the Fowler brothers insisted that anyone can improve their lot in life once they understand how their mind is constructed. The motto they chose for their masthead—"Know thyself"—suggested that understanding the shape of your brain was merely the first step on a long path toward self-realization.

Whitman bought into all of it, and in his poems he created the idealized image of Walt Whitman of the body electric, who had been endowed with beautiful blood and a beautiful brain. And yet, despite his glowing phrenological diagnosis, he knew better. He was aware that chronic illness ran in the Whitman family. His parents suffered from a litany of physical complaints, including dizziness and headaches. So did he. Neurological disease had been passed down to him through several generations. His father, Walter Whitman, had died of a cerebral hemorrhage after suffering years of paralysis on one side of his body, the result of a stroke. His brother Jeff exhibited the facial asymmetry typical of stroke victims. Walt himself, despite appearing quite robust to Lorenzo Fowler, struck others as being somewhat sluggish and heavy-limbed, something of a laggard. His perfect body and perfect blood line, historians now concede, was a myth the poet crafted and perpetuated for the sake of his poetry long after the failure of that perfect body was plain to see. In a sense, phrenology held out his only hope.

Whitman's stint as a volunteer nurse during the Civil War is well known. What is not so well known is the toll it took on his nervous system. When the war broke out in 1861, Whitman was too old to fight, but his brother George enlisted at the first opportunity, fought at Roanoke, at Second Bull Run, then at Fredericksburg, where a minié ball grazed his cheek and sent him to the hospital.

While scanning the lists of wounded in the newspapers, Walt came across George's name, and instantly set out for Virginia to find him. To his great relief, he learned that his brother was not seriously injured, but others were. For nine days he stayed in the camp, tended to soldiers, recorded their stories, and lifted their spirits. On his way home he got as far as Washington, where he took a job at the army postmaster's office. In his spare time he visited hospitals, volunteered his time and energy, and in the process sacrificed his already compromised health. The stress was enormous, and it brought on chronic headaches and dizziness, the very symptoms that had plagued his parents. In the wards he contracted miliary tuberculosis, one of many physical problems that would plague him, without killing him, for the rest of his life. (Signs of it would show up in his autopsy some three decades later.) It was also where he developed attachments to young men, and the anxiety brought on by his emotional confusion, so biographers have claimed, would lead to a catastrophe.

After the war, Whitman remained in Washington and worked at the Department of the Interior, then at the attorney general's office. His "brain troubles," as he called them, continued to plague him for the better part of a decade, culminating in 1873 in a crippling stroke that left him partially paralyzed. Although his mind was unaffected (he could still write), he was unable to work. Only fifty-three years old, and facing the possibility of not being able to take care of himself, Whitman had no choice but to claim his meager pension and move to his brother George's house in Camden, New Jersey.

It turned out to be a good choice. In the 1870s, the booming city across the river from Camden was the leading scientific city in America, and home to its most celebrated anatomists, surgeons, physicians, scientific societies, and medical schools. Philadelphia was also, due to the presence of two men—Silas Weir Mitchell and William Osler—the nation's center of neurology, experimental physiology, and clinical medicine. As luck would have it, Whitman would soon enjoy the attentions of both of them. All it would cost him was his brain.

IT IS NOT CLEAR who induced Walt Whitman to join the American Anthropometric Society, or whether he officially joined at all.

What is known is that the society claimed his brain when he died in 1892, and that at least three, and possibly four, of his physicians during his Camden years were brain men. One of them, easily the most influential, was Silas Weir Mitchell.

It is hard to sum up Silas Weir Mitchell in a few sentences, although Whitman came close. "He is of an intellectual type," he once remarked, "a scholar, writer, and all that: very good—an adept: very important in his sphere—a little bitter I should say—a little bitter—touched just a touch by the frosts of culture, society, worldliness—as how few are not!"[7]

Mitchell was indeed "very important in his sphere." Through sheer determination, he had become the country's leading experimental physiologist, but that gives too narrow a measure of the man. He was also a successful poet, novelist, and essayist, which would mean more if any of his books, poems, or essays were still read, or even remembered. Judged by readership and earnings alone, he was a far more accomplished writer during his lifetime than Whitman could ever have hoped to be. His most successful novel, a historical romance called *Hugh Wynne*, sold over a half million copies. It was of this book that Mitchell once said, pondering his legacy, that it might enjoy "the notoriety of a decade."[8] False modesty, perhaps, but as it turned out, right on the mark.

In the 1880s, Silas Weir Mitchell was known around Philadelphia as the most eminent of eminent men, a social paragon, a man of the sciences, arts, and public affairs. He was, in effect, the heir apparent to Benjamin Franklin as leading light of the city, a Renaissance man without whose presence no social event could be considered a success. Yet Whitman could see, beneath the refined manners and the patrician demeanor, the seeds of growing bitterness and frustration. Mitchell had been passed over for several university professorships because he was an experimentalist. Although he was the most important neurologist in America, he could not land a steady job. Like many researchers, he had been obliged to hang out a shingle and consult at hospitals. Even so, he succeeded in changing the course of medical education in the United States. During the late 1870s and early 1880s, he even found the time to serve as Walt Whitman's personal physician.

Mitchell was born in Philadelphia in 1829. His father, John Kearsley Mitchell, one of the city's most distinguished physicians, was a member of the first phrenological society in America, and a devoted follower of phreno-magnetism (a form of therapeutic hypnotism invented by the Austrian physician Franz Anton Mesmer in the late 1700s). He did his best to discourage young Weir from following him into the profession, but to no avail. Weir Mitchell received his M.D. from Jefferson Medical College (where his father taught) in 1858, made an obligatory and undistinguished tour of European research laboratories, and returned to Philadelphia intent on becoming a physiologist. His father took a dim view of the plan, brought his son into his own practice, and loaded him up with work. But Mitchell persevered. He had just begun to specialize in studies of the nervous system when the Civil War broke out.

Rather than enlist, Mitchell contracted himself to the army as a surgeon, and took charge of a new four-hundred-bed hospital in Philadelphia. He and his colleague, William W. Keen, began to investigate cases of battle fatigue and neuralgia (a term used to describe shooting pains that traverse the nerve paths out to the limbs). They were especially interested in soldiers who reported symptoms that had no organic explanation, such as paralyses induced by incidental wounds or by emotional trauma, an effect called reflex paralysis. Mitchell also became an expert at detecting malingerers. He traveled to Gettysburg after the last day of the battle and brought back wagonloads of injured soldiers. From their case studies, Mitchell and Keen produced two landmark publications: *Gunshot Wounds and Other Injuries of the Nerves* and *Reflex Paralysis*, both published in 1864. When they were reprinted eighty years later by the Yale Medical School, the editor called them "one of the great milestones in the history of American neurology and American clinical medicine."[9] *Gun-*

shot Wounds, in particular, remained the authority on its subject for several decades, and was still in use by the French army as late as World War I.

Mitchell was not a brain specialist. His investigations into brain anatomy were limited to the physiology of the cerebellum, and are not well remembered. On the other hand, his descriptions of symptoms of soldiers whose limbs were amputated are considered classic works of clinical diagnosis. He coined the phrase *phantom limb syndrome* to describe reports of sensations in missing arms and legs (a finding so controversial that he published it under a pseudonym in order to avoid ridicule).

By the end of the war, Mitchell had joined the ranks of the world's elite researchers, yet he still could not land a professorship. Despite recommendations from the foremost scientists in the country, he was passed over at both the University of Pennsylvania and at Jefferson Medical College in favor of less-qualified, but more traditionally credentialed applicants. It was the greatest source of frustration and bitterness in his professional life, but within a decade it would cease to matter.

In 1875, a dozen years after the death of his first wife, Mitchell married into one of the wealthiest families in Philadelphia, the Cadwaladers. He would never again have to worry about money, nor would he have to strive for social status. He summered at Newport, and so boring did he find this obligation that he began to pass the time by writing novels. For years he had been writing poetry and stories, but Oliver Wendell Holmes, the great poet-physician, had warned him not to publish until he had established his medical practice. People tend to be suspicious of doctors who write, said Holmes. As a result, Mitchell did not publish his first novel until 1880, when he was fifty. It was an instant success.

When Mitchell is remembered today, it is not for his poetry or for his prose or for his many physiological discoveries, but for his greatest therapeutic breakthrough, the rest cure, a treatment he invented after years of dealing with neurotic women. Although scoffed at now, the method had its adherents, and Mitchell could point to scores of successes, among them the novelist Edith Wharton, his most famous patient. In *The Wound and the Bow*, the critic Edmund Wilson claimed (without much evidence) that Wharton began writing seriously at Mitchell's suggestion as a way of recovering from her nervous breakdown.[10]

The rest cure is exactly what its name implies—a complete dropping out from everything, being confined to bed, and following the blandest of diets. Mitchell claimed to have come up with the idea while treating soldiers suffering from acute exhaustion, a condition that later came to be called shell shock. The treatment could last for months, and it was infamous for its properties of transference. Nurses routinely fell ill out of sympathy, if not sheer exasperation. "I have seen a hysterical, anaemic girl kill in this way three generations of nurses," Mitchell remarked.[11]

Mitchell cut a tall, lithe, and dapper figure. Women found him dazzling, and had he been so inclined, he could have had his pick of them. But he was not so inclined, and it was this very restraint, some literary critics contend, that weakened his novels, that emasculated his poetry, that brought him to the threshold of delving into the sexual undercurrents of bourgeois society, yet prevented him from stepping into the bedroom.

He certainly had stories to tell, most of which went with him to the grave. A suggestion of his ease with women survives in the anecdotes he enjoyed telling at gatherings. Perhaps the most famous concerned a woman who refused to end her rest cure, who would not get out of bed. Mitchell dismissed the nurse from the room, and said, "Madam, if you are not out of bed in five minutes—I'll get into it with you!"[12] And he started to undress. Fortunately (or perhaps unfortunately), the gambit worked before he had his trousers completely off. In a similar case, he sent the nurses away and told them to close the door behind them. A few minutes later he emerged, quite calm, and declared, "She will be running out of the room inside of two minutes." Seeing the dubious looks around him, he added, "I set her bed sheets on fire. A case of hysteria."[13]

Silas Weir Mitchell was a vain man, even more so than Whitman, and he did little to hide it. When chastised by his daughter-in-law, he simply replied, "My dear, I have something to be vain of!" But Whitman saw through the façade.

"It is true that Mitchell has written poems," Whitman said one day to his friend, Horace Traubel, "a volume at least or two—I am moved to second you when you say that they don't come to much (I guess they don't)—they are non-vital, are stiff at the knees, don't float along freely with the fundamental currents of life, passion. But then you know that in our time every fellow must write poems—a volume at least—and a novel or two—otherwise he can't qualify for society: he writes, he writes, then he gets over it all—recovers."[14]

In a more generous spirit, Whitman once said of him, "I don't know but he's about as near right in most things as most people. I can't say that he's a world author—he don't hit me for that size—but he's a world doctor for sure—leastwise everybody says so, and I join in."[15]

If Mitchell had been capable of such bluntness, he refrained, out of a sense of decorum, from working it into his prose. He was an aspiring artist of some talent who was undone by his overbearing sense of propriety, by the narrow confines of his social class, by his isolation from new ideas.

"What kept Weir Mitchell from being first rate?" his own biographer, Ernest Earnest, was moved to ask.[16] An accident of birth. Mitchell, he wrote, had the misfortune to be born in the most self-effacing city on the planet—that bastion of conformity, complacency, and class consciousness that is Philadelphia. "He once boasted," wrote Earnest, "that he never wrote a line which might cause a young girl to blush." It was a problem that Walt Whitman never had.

IN 1884, WHEN William Pepper vacated his chair of clinical medicine at the University of Pennsylvania in order to become the university's provost, an important opportunity presented itself. Like his counterpart, Charles Eliot of Harvard (who was a chemist before taking over the presidency), Pepper was intent on reforming American higher education by building the reputation of its professional schools, and by establishing research as a priority on a par with teaching. He would be aided considerably in this goal by Silas Weir Mitchell, who had become a university trustee, and who personally set out to find the most qualified man to fill Pepper's place. In other words, he set out to find a younger version of himself. The man he found was William Osler, a pathologist at McGill University in Montreal.

A generation earlier, the job would have gone to a clinician with social connections and no research background. But Osler had an impressive record of publications to supplement his hospital work (he had been a pathologist at Montreal General), and was revered at McGill as an inspiring teacher. Osler belonged to a new breed of professors. Unlike the stodgy lecturers of old, who left all demonstrations to their assistants, Osler insisted on hands-on learning, and he taught medical science as a laboratory science. One of his primary research interests, not surprisingly, was brains.

Osler arrived in Philadelphia in 1884, and had barely settled in when he received a letter from a friend in Quebec asking him to "look in on old Walt if you have a chance."[17] The friend was Richard Maurice Bucke, a Canadian physician who also happened to be one of Whitman's confidants. Although he had never heard of Walt Whitman or read a single one of his poems, Osler paid a visit to Camden. He had no idea that Whitman was the patient of the man who had brought him to Philadelphia.

The meeting went well. Whitman sized up Osler as he sized up everyone—at a glance—and liked what he saw. Osler was also impressed, and he offered his services at no charge. Silas Weir Mitchell had become too busy to visit Whitman in Camden, and Walt could no longer easily make the trip to Philadelphia, so the arrangement worked out perfectly.

Osler was on very familiar terms with Whitman, much more so than Mitchell had been. Few of their conversations were recorded, but they must have discussed phrenology at some point. Osler dabbled in it, like most physicians of the time, and had noted its failings, but he remained intrigued by the possibilities of cerebral localization. Whitman, on the other hand, had a nostalgic affection for it. Sometime in 1888, he admitted, "I know what [Oliver Wendell] Holmes said about phrenology—that you might as easily tell how much money is in a safe feeling the knob on the door as tell how much brain a man has by feeling bumps on his head: and I guess most of my friends distrust it—but then you see I am very old fashioned—I probably have not got by the phrenology stage yet."[18]

Osler treated Whitman from 1885 to 1889, then left for Baltimore to head up a new medical school at Johns Hopkins University, which he soon built into a world-class research institute. He was living in Baltimore when his former colleagues met to form the American Anthropometric Society. Like Mitchell, Osler could not attend the first meeting, but signed on at the first opportunity. He returned to Philadelphia periodically to take part in the social occasions that doubled as meetings, and during those visits he tried to persuade Whitman to come to Baltimore with him, to move into a facility on the hospital grounds, where he would be assured of round-the-clock care. But Whitman declined. He preferred his simple routine and the company of his friends at the little wood-frame house on Mickle Street in Camden. He intended to die there.

ALTHOUGH WALT WHITMAN'S demise had seemed imminent on many occasions, he began to fail noticeably in November of 1891. Osler and Mitchell were no longer on call, and his doctors at that time, Daniel Longaker and Alexander MacAlister, had no real idea what was wrong with him. Whitman himself did little to help them. He rarely complained. In December, however, pneumonia set in, he became feverish, and his left lung collapsed.

Three months later, on the evening of March 26, Whitman died in the presence of his inner circle—Dr. MacAlister; his housekeeper Mrs. Davis; Warren Fritzinger, his nurse; Thomas Harned, his attorney; and Horace Traubel, his longtime friend. Daniel Longaker was sent for, but he did not arrive from Philadelphia in time. Fritzinger shifted Whitman some sixty-five times that final day, providing what little comfort he could, and Whitman's last words, barely strung together, were, "Warrie. Shift."[19]

Thomas Eakins, who had painted Whitman's portrait four years earlier, arrived the next morning with his assistants to make a death mask and casts of Whitman's hands. Meanwhile Horace Traubel went to the university to confer with Harrison Allen and Francis Dercum about the autopsy.[20] It seems that sometime during the previous months, possibly during the crisis of December, Whitman had agreed to a postmortem examination. "If it will be of interest to the doctors," he is alleged to have said, "and of any benefit to medical sciences, I am willing."[21] The source of this remark was Daniel Longaker, who, as a Penn medical school graduate and a friend of Allen's and Dercum's, was probably a member of the AAS. What makes the connection seem even more likely is the fact that Longaker raised the issue with Whitman in Harrison Allen's presence (Allen was at the time the Brain Society's president). Allen was then able to confirm the story for Traubel.

It is not clear whether Whitman willed his brain to the society, but it seems likely that he would have been happy to add his name to the list. Although few of his private conversations with Mitchell or Osler were recorded, they probably discussed brains and the current state of research on cerebral localization. There is no doubt that Whitman knew of the Brain Society's existence. Not only was he acquainted with several of its principal members, but he had read and remarked upon the deaths of Joseph Leidy and Philip Leidy the previous year. He knew precisely what had happened to their brains.

Joseph Leidy died unexpectedly in April of 1891. Harrison Allen performed the autopsy and removed the brain, with Francis Dercum assisting. By an odd coincidence, Leidy's brother Philip had died on the previous day, and Dercum had removed that brain, with Allen assisting. Both deaths were well publicized, and reporters played up the connection to the AAS, giving details of the cursory examination of the brains and the plans for their further study.

The sudden deaths of the Leidy brothers highlighted a thorny problem for the brain men. Someone had to be ready to go out at any time of day, in any conditions, to acquire the brains of deceased society members. The task did not end there. There was also the time-consuming process of preserving the brains as they accumulated. In addition, the plan called for plaster casts to be made and distributed to the membership. If the collection and storage of the brains was going to be done right, they would have to hire someone for the job.

Fortunately, a qualified candidate was already on hand—Henry Ware Cattell, an 1887 graduate of the University of Pennsylvania School of Medicine, and a demonstrator of morbid anatomy. Cattell accepted the position of prosector for the society in 1891, and he was initiated into the role by assisting Dercum at the autopsy of Andrew J. Parker, a professor of biology, who died in March of 1892. Within a week, the two men would set off again, this time in the direction of Whitman's house in Camden.

A PROSECTOR IS someone who wields the knife at an autopsy or a classroom dissection under the watchful eye of the supervising pathologist. At Whitman's autopsy, Dercum is said to have removed the brain, but most of the routine work would have been performed by Cattell, who was an accomplished pathologist in his own right. In 1894, he would write a textbook on the subject of postmortem examinations, and Whitman's autopsy would have been fresh in his mind when he addressed the delicate subject of permission.[22]

Whatever Whitman might have said privately to the brain men, he did not make his wishes known to his brother. Which is to say that the American Anthropometric Society did not have permission to perform a postmortem, much less George Whitman's blessing to remove and take away his brother's brain. George adamantly declared his opposition to the procedure, but he would make two crucial mistakes. The first was to say that if any scientific end would be served, he would have no objection. What he added, but did not emphasize enough, was that as far as he was concerned an autopsy would merely satisfy the physicians' morbid curiosity. His second mistake was to go home before the matter had been resolved. In his absence, Dercum and Cattell waited for Eakins to finish the death mask, then held a brief meeting in which they decided that the interests of science would indeed be served.[23] They also decided that it would not be necessary to tell George Whitman, and they went ahead with the autopsy.

As they worked their way through the roster of organs, each one ravaged seemingly beyond the point of sustaining life, the physicians grew more and more impressed. Whitman's left lung had collapsed and was useless. Only one-eighth of the right lung was functional; the rest of it was consumed by tubercular nodules. Signs of emphysema and pneumonia were everywhere. Tubercles dating back perhaps thirty years had compromised his stomach, kidneys, liver, and spleen. Whitman had been reduced to the brink of failure in every organ but his heart and brain. The official cause of death was listed as pleurisy of the left side, consumption of the right lung, general miliary tuberculosis, and parenchymatous nephritis (essentially kidney failure).

As Francis Dercum worked at removing the brain, Whitman's good friend Horace Traubel looked on and wrote down Cattell's notes as they were dictated. "To hear the claw and dip of the instruments,"

Traubel would later reflect, "to see the skull broken and opened and the body given the ravening prey of investigator had its horrors, then its compensations."[24]

Dercum commented on the "magnificent symmetry of the skull," and recorded a brain weight of 45 ounces (1,276 grams). He probably did not weigh the specimen at the scene. George Whitman's disapproval hung like a cloud over the proceedings, which seemed to be marked by haste. According to Traubel, "When the brain was extracted, Cattell put it into his gupsack." (Although not found in dictionaries of the era, the word evidently means an oil-cloth bag.) Dercum and Cattell then left with their fragile specimen.

The autopsy report, written in Cattell's own hand and published in the commemorative volume, *In Re Walt Whitman,* that appeared in 1893, stated very plainly:

> The brain was removed by Dr. Dercum, and is now, after having been hardened, in the possession of the American Anthropometric Society. This Society, which has been organized for the express purpose of studying high-type brains, intends to first photograph the external surfaces and then make a cast of the entire brain. After this, careful microscopic observations will be made by competent observers.[25]

But this never happened. With one significant exception, Whitman's brain would not be heard of again for the next fourteen years.

WHEN WILLIAM PEPPER died of a heart attack in California in the fall of 1898, the American Anthropometric Society lost its founder and third president. Pepper had been warned about overwork but had kept to his busy schedule. Like most of his colleagues, he was incapable of resting. His assistant, Dr. A. E. Taylor, removed Pepper's brain and brought it back to Philadelphia, where it was placed in a vault in the basement of the Wistar Institute.

Ever since Whitman's death six years earlier, the brain men had kept a low profile. But the death of Cope in 1897, and now that of Pepper, served to remind reporters that a brain-donation society still existed, that it was still worth a half-page feature. After the reading of Pepper's will, the *New York Herald* sent a reporter to Philadelphia to investigate, and the story he filed ran under the headline "Three Hundred Men Pledged Their Brains to Science," with a subhead that added, somewhat less credibly, "Seventy Have Already Died; and Their Gray Matter Reposes in the Archives of the Wistar Museum in Philadelphia."[26] The figure is not correct. There were never more than two dozen elite brains on the premises (there are twenty-two in the collection at present). As of 1898 there were only seven. According to the institute's records, the other brains did not belong to the American Anthropometric Society, but came mostly from the Elwyn and Vineland schools for the mentally retarded. Someone, it seems, had been feeding tall tales to the press.

The "someone" in this case, the reporter's lone source for the story, was Henry Cattell, the brain society's prosector, who had good reason to be evasive. Cattell, it seems, had managed to lose Whitman's brain, and he did not want the fact revealed. It would have been embarrassing, if not disastrous for the Brain Society, to say that he had accidentally destroyed it (which is what seems to have happened). The simplest way out, he decided, was to tell the reporter that he had never had it.

"Dr. Cattell had made an effort to secure the brain of Walt Whitman," the story said, "but had encountered the vigorous opposition of the venerable bard's family. Nevertheless he made the autopsy on March 27, 1892, and weighed and examined the brain, although he was not permitted to take away the tissue." Had anyone at the newspaper checked the files, they would have known that this was not true.

Cattell's own autopsy report contradicted it, as did eyewitness testimony.

The only logical explanation for the odd series of events surrounding the fate of Whitman's brain is this: Cattell must have written the autopsy report shortly after Whitman's death, anticipating what *would have been* the case when the report was eventually published. (Unfortunately, the report is not dated.) At some point it seems, the brain was destroyed, probably within days of the autopsy. It could have happened in any number of ways (the very word *gupsack* suggests all sorts of nasty possibilities), but it was probably dropped in the lab as legend has it, perhaps by Cattell, or destroyed during the risky process of making a cast of the hemispheres. Had it survived the hardening process, as any anatomist knows, it would not have broken into pieces when dropped unless it had been improperly preserved.

Cattell was clearly covering for himself. His inflated estimate of the number of elite brains, another clever detail, made the absence of one poet's brain seem inconsequential. No one, he naïvely assumed, would notice the discrepancy. But as it would turn out, Cattell was sadly mistaken.

Chapter 12

Spitzka

THE LOSS OF Walt Whitman's brain might have gone unnoticed if it had not been for a resurgence of interest in the American Anthropometric Society instigated by Edward Charles Spitzka's son, Edward Anthony Spitzka. In the decade that had elapsed since the founding of the society, the elder Spitzka had drifted away from his friends in the Philadelphia group. He still lived in New York, still ran a private neurological practice out of his town house on East Seventy-third Street, and had kept up a steady output of neuroanatomical research, mostly on animals but occasionally on human brains. (His name entered the anatomy textbooks with his discovery of the "marginal tract.") Although he had collected several interesting specimens over the years, notably the brains of the Séguins, father and son, Spitzka had not found the time to study them. But he was gratified to find out that his son was willing to take on the job.

During the 1890s, Spitzka's growing brain collection must have been a source of increasing wonder to young Edward Anthony Spitzka, who followed in his father's footsteps by attending P.S. 35, then City College, and finally the College of Physicians and Surgeons at Columbia University, where he distinguished himself as the best anatomist in his class. At the age of twenty-four, in only his second year of medical school, Spitzka had already chosen a research specialty. He decided to study exceptional brains, starting with his father's neglected collection.

With the departure of Henry Cattell in 1899, the American Anthropometric Society had lost its only active member. During his eight-year tenure as the society's prosector, Cattell had difficulty following through with the mission. He did not produce a single brain study, nor did he keep any records of the group's efforts. He managed to collect several important brains, but instead of "archiving" them and making plaster casts, he stored them at various locations without informing the membership of their whereabouts. Most of them were eventually consigned to the Wistar vault.

When Edward Charles Spitzka, who had been out of touch with his fellow brain men for a decade, made some inquiries on his son's behalf, he discovered that the AAS had a new president, Joseph Leidy Jr., who bore the name of his famous uncle, but was in fact the son of Philip Leidy. In January of 1901, Spitzka's son dutifully sent off a letter to Leidy, offering his services. "It seems to me that a delay in examining and describing such illustrious men's brains almost borders on disloyalty," he wrote. "If what is needed is someone enthusiastic enough to devote much time and labor to such tasks . . . you have only to command me, and as soon as my college examinations are finished, I would deem it an honor to do so."[1]

Leidy replied that the society would be happy to make the brains available on one condition: they could not be dissected. Another stipulation, stated more as a request, was that Spitzka agree to assume Cattell's role as the society's prosector. At the time, Spitzka appeared to be the best candidate for the job. He had the expertise, energy, interest, and the time to fulfill the Brain Society's mission. He was also the only candidate: No one else was willing to do it. So a deal was struck. As his first order of business, Spitzka set about making an inventory of the brains.

Nineteen hundred one would be a breakthrough year for Edward A. Spitzka. Before it was over, he would join the inner circle of the

Brain Society, publish his first studies of special brains, and start a brain collection of his own. If the parallels with his father's life do not seem uncanny enough, he would add one other unusual accomplishment to his résumé when he dissected the brain of a presidential assassin in the wake of a national tragedy.

ON SEPTEMBER 6, 1901, a self-professed anarchist named Leon Czolgosz stood patiently in a receiving line at the Pan-American Exposition in Buffalo, New York, gripping in his right hand a pistol concealed under a handkerchief. Upon reaching the front of the line, he greeted the outstretched hand of the president of the United States with two gunshots. William McKinley died eight days later, and, in an astonishingly efficient execution of justice (or vigilantism), Czolgosz was sentenced, tried, and convicted, all within the next twelve days. Less than two months after firing the fatal shot, Leon Czolgosz was put to death in the electric chair. It was Spitzka, just twenty-five years old, and still Mr. and not Dr. Spitzka, who performed the postmortem, removed Czolgosz's brain, and wrote the autopsy report.

It was an astounding feat, and a galling one in some quarters—the murderer of a president examined by a mere student. But even as a student, Spitzka had amassed some impressive credentials. With eight journal articles to his credit, he had already been accepted for membership in the prestigious American Association of Anatomists. No doubt the Czolgosz assignment would not have happened without connections. Carlos MacDonald, the New York State medical examiner, was a close friend of Spitzka's father as far back as the botched execution of George Kemmler in 1889. Because the Czolgosz autopsy would require a brain expert, MacDonald probably asked the father to attend the execution, but instead he got the son. It turned out to be a good choice.

Spitzka's postmortem on the brain of Leon Czolgosz was published simultaneously in six medical journals. It was unanimously judged a first-rate piece of work. Knowing that theories of physical stigmata dominated the thinking of contemporary criminologists, Spitzka limited his conclusions to the objective neutrality of anatomy.

"Nothing has been found in the brain of this assassin," he wrote, "that would condone his crime for the reason of mental disease due to intrinsic cerebral defect."[2] As his lone speculation, Spitzka appended a prescient comment, one that contradicted his father's theory that almost all psychoses result from lesions of the nervous system: "It is well known that some forms of psychoses have absolutely no ascertainable anatomical basis, and the assumption has been made that these psychoses depend rather upon circulatory and chemical disturbances." Six years before the theory of neurochemical mediation of cell function was first established, Spitzka seems to have intuited the principal direction that neuroscience would take during the twentieth century. Unfortunately, he then set off in the opposite direction.

EVERYTHING FELL INTO place for Spitzka after the Czolgosz autopsy. Thanks to the efforts of his father's generation, he did not have to go to Europe to learn a medical specialty. By 1900, the United States possessed some of the premier medical research institutes in the world, which allowed Spitzka to hone his skills first at Columbia, where he stayed on as a fellow in anatomy for two years under the eminent pathologist George Huntington (who first described the degenerative nerve disorder called Huntington's chorea), then at Jefferson Medical College in Philadelphia, as a demonstrator of anatomy under the surgeon J. Chalmers DaCosta, the editor of the American edition of *Gray's Anatomy*. During this time Spitzka published a string of important studies that solidified his reputation as an up-and-coming anatomist. His stock rose so high that in 1906, just four years out of medical school, he was offered a chair in general anatomy at Jefferson, and his future seemed assured. DaCosta chose him to coedit the seventeenth American edition of *Gray's Anatomy*, and Spitzka would become its sole editor for two subsequent editions.

In addition to his teaching and laboratory work, Spitzka was a visiting physician at local prisons, a job that came his way after the Czolgosz autopsy. In a scene he reenacted dozens of times during those years at any prison within a day's travel, Spitzka would arrive with his small satchel of tools, he would examine the condemned man, witness the execution, perform the autopsy, harvest the brain, and cart it back to his lab. He soon became an authority not only on criminal brains, but on the effects of electrocution on brain tissue. In the process he became an advocate for electrocution as a humane alternative to execution by hanging.

While building his sizable collection of criminal brains, Spitzka also immersed himself in the examination of brains of eminent men. Drawing first upon his father's collection, he wrote studies of the brains of Édouard and Edward Séguin, and of the brains of three diabolical brothers, convicted of a single murder, who were executed on the same day at New York's Dannemora prison. With the field almost to himself, Spitzka easily acquired other highly prized specimens, including the brains of the anthropologist and explorer John Wesley Powell, and of the merchant George Francis Train, each of which led to a lengthy monograph. He also managed to acquire the brains of four Greenland Eskimos who had been brought to New York in 1897 by Admiral Robert Peary, and housed in the basement of New York's Museum of Natural History. Of the six men Peary procured for the museum, four soon died and were dissected. (The anthropologist Franz Boas staged a mock funeral for one of them in order to appease the man's father, then proceeded with the autopsy. The museum displayed the skeleton; Spitzka studied the brain). But the most important brains on Spitzka's agenda belonged to the brain men—Leidy, Cope, Pepper, and the rest—and they came to his examining table one by one as the talented anatomist prepared to undertake the most ambitious comparative brain study ever made.

When he made his first inventory of the Wistar collection in 1902, Spitzka was surprised at the deplorable state of the specimens. Out of the ten brains that had been catalogued, only seven had been deposited in the Wistar vault. Of these, four had been damaged through neglect or mishandling, while two others had been improperly weighed. It was a pretty sorry record. In a preliminary report to the membership in 1906, Spitzka felt obliged to enumerate these setbacks.[3]

The brain of Dr. Andrew J. Parker, he noted, had been left in Müller's fluid, a hardening agent, and had crumbled to pieces. Fortunately a plaster cast had been made, allowing it to be included in the study. The brain of Professor J. William White, on the other hand, had suffered irreparable damage, and no cast existed. A total loss. The brains of Harrison Allen and William Pepper had become flattened in their jars, yet retained enough of their shape to be of use. The weighing of Joseph Leidy's brain had been botched (Harrison Allen had apparently been flustered at the prospect of carving up his mentor), and the fresh brain weight could only be estimated. Three brains were missing. Those of Isaac Newton Kerlin and his wife, Harriet, were said to be in Henry Cattell's possession and were not made available. Spitzka declined to say why. The other specimen that should have been in Cattell's care was the brain of Walt Whitman.

The loss of Whitman's brain, as far as Spitzka was concerned, was unfortunate but not catastrophic. He was told (by whom it is not clear) that the jar containing it had been dropped, probably by a laboratory assistant, and that the broken pieces had been discarded. Spitzka did not think to question the story or to investigate it. He had six other brains to study, and that would have to suffice.

In addition to examining the brains, Spitzka delved into the historical literature on the subject, particularly the work of Rudolf Wagner and the French anthropologists. In order to place his own work in its proper context, he pieced together a history of elite and criminal brain investigations going as far back as Byron. After five years of intense labor, he was at last ready to release his findings.

On December 5, 1907, the *Philadelphia North American* broke the story of Spitzka's research under a banner that may have been intentionally ambiguous: "Brain Research by Phila. Anatomist Startles Science!"[4] It was either the scientific breakthrough of the year, or just another false alarm. In either case, the article suggested, it was "certain to create wide discussion and general interest," which it did. To Spitzka's chagrin, most of it would have nothing to do with science, and almost everything to do with a throwaway comment that he had inserted on page one, in which he revealed the sad fate of Whitman's brain, how it had slipped out of the hands of a laboratory assistant, had broken into pieces, and been discarded. Had he any idea what would come of that comment, he probably would have kept his mouth shut.

On the following day, the *North American* ran two related stories. One was to be expected: "Scientists Attack Dr. Spitzka's Brain Structure Theory," it said. The other was not. It was headed: "Loss of Poet's Brain Roils His Executors."[5]

Instead of responding to the technical points Spitzka had raised in his monograph, the press began to besiege him with questions about the dead poet. Who was responsible? How could it have been allowed to happen? The press lined up a gallery of likely suspects, including the Wistar Institute, Jefferson Medical College, and even Spitzka himself. When pressed by reporters, he snapped, "No blame can attach to me. I had nothing to do with its keeping. I decline to discuss what the executors of the poet may say." But he must have experienced a sinking feeling that his magnum opus was dead on arrival, while the story of the shattered brain had taken on a life of its own.

SPITZKA'S "A STUDY of the Brains of Six Scholars and Scientists," fascinating as it is both as literature and as history, fails completely as science. It delivers none of the breakthroughs promised by its author. Even so, its opening section is a tour de force. In it, Spitzka catalogues 137 case studies of brains of accomplished men and women, followed by eight "doubtful reports" (including those of Oliver Cromwell, Lord Byron, Franz Schubert, and Blaise Pascal). Most of the entries are perfunctory, and give little more than a name, dates, brain weight, and a few comments, as in:

> Paul Broca (1824–1880), French Anthropologist (Paris collection). The brain was weighed by M. Kuhff. The brain weight was 1484 grams. No further records seem to have been made of this brain.

Other entries contain some intriguing trivia. Oliver Cromwell, for example, must have had at least three heads, given that three different collections claimed to possess his skull. His brain, according to entirely unreliable accounts, was huge (as was Pascal's, apparently). Officially, the heaviest elite brain of all, weighing in at an astounding 2,012 grams, belonged to the Russian novelist Ivan Turgenev. The great orator and statesman Leon Gambetta had the smallest, at 1,160 grams. Abraham Lincoln's brain was removed in order to find the track of the bullet that killed him. It was weighed piecemeal, and because of the loss of tissue the result was not accurate, but it placed him in a range "not above the ordinary for a man of Lincoln's size."

Many entries cry out for further explanation. In his profile of Wilhelm Steinitz, winner of the first world chess championship held in 1866, Spitzka recounts Steinitz's descent into melancholia and madness by quoting from L. C. Petit's *The Pathology of Insanity:* "He spent much time gazing into space 'trying to hypnotize Bab the Persian God.' From a partially systematized insanity he soon became overwhelmed with delusions of persecutions and hallucinations." He describes the brain as "almost phenomenal in the development of the orbital and frontal convolutions," adding that Steinitz was a small man with an unusually high brain-to-body weight ratio. He leaves it to the reader to judge the significance.

Ambiguity was Spitzka's signature. In describing a brain, particularly its convolutions, he relied on expressions such as "splendidly developed," "well developed," "richly convoluted," and "quite sinuous." He often spoke of "doubling" or "redundancy" in a given lobe, to indicate a surplus of brain matter (presumably in the range of a 6 on the phrenological scale). In his catalogue of famous brains, that of the Russian general Mikhail Skobelev even merits the adjective *bombifrons,* a word not to be found in any dictionary, although its meaning seems clear enough.

In listing the case histories (in chronological order by year of death), Spitzka hints at the self-devouring nature of the whole enterprise. The same investigators appear in bibliographical references over and over, only to show up for the last time as entries themselves. The brain of the master dissected by the assistant, whose own brain then shows up a few pages later, in a big-fish-eats-small-fish progression. The same fate awaits me, Spitzka seems to be saying with stoic indifference. ("To me," he writes, "the thought of an autopsy is certainly less repugnant than I imagine the process of cadaveric decomposition in the grave to be.")

Most of the entries were culled from autopsies. It was not unusual for a lone investigator to acquire a single specimen in order to satisfy his curiosity or indulge the ego of a colleague. Spitzka records several instances of distinguished men who jumped at the chance for a postmortem confirmation of greatness. The English historian George Grote, for example, left detailed instructions for the disposition of his brain, resulting in two lengthy monographs by the physician John Marshall. The brain of Charles Babbage, the mathematician and inventor of the difference engine, an early computer, received similar treatment.

Spitzka takes pains to note which brains can still be found, as of 1906, in existing collections—the Göttingen collection, the Munich collection, the London, Paris, Stockholm, Turin, Philadelphia, Cornell, Moscow, or "author's personal" collections. In a short preface he traces some of their histories. Göttingen, of course, means Rudolf Wagner; Paris means the Société Mutuelle d'Autopsie; Cornell means Burt Green Wilder and his Cornell Brain Association; Turin refers to the laboratories of Cesare Lombroso and Carlo Giacomini; Stockholm refers to the brilliant anatomist Gustav Retzius and a colleague named Tigerstedt, who tried to start a brain-donation society at the prestigious Karolinska Institute in the late 1890s. They circulated a petition among their colleagues, and, after it had made the rounds, it was discovered to contain only two signatures: their own. This initial setback did not prevent Retzius from acquiring the brains of several prominent Karolinska scholars, including the Russian mathematician Sonya Kovalevsky, and the astronomer-mathematicians Hugo Gyldén and Per Adam Syljestrom. (Retzius's own brain can be found today at the institute, along with that of the Swedish poet Gustav Fröding.)

Like his father, Spitzka believed in Francis Galton's positive eugenics, in the importance of propagating the right stock. In his view, the United States, with its influx of Europeans, was accomplishing the task quite naturally. "Nowhere in the world," he wrote, "is the mixture of races—chiefly the Teutonic, Celto-Roman, and Slavonic—going on so actively as in this country."[6] Another of Spitzka's inherited axioms was the superiority of northern races over southern ones. In his view, North America presented conditions that were "peculiarly advantageous to the preservation and restoration of the best types," and he predicted future generations blessed with bigger and better brains. As for "the tropical races," he wrote: "living at ease, amid great plenty of food and with much leisure time for quarreling amongst themselves, the story is quite different."

IN STEPHEN JAY GOULD'S *The Mismeasure of Man*, Spitzka comes in for rough treatment because of his casual remarks about race. In his 1903 monograph on the brain of John Wesley Powell, for example, Spitzka was playing to the crowd when he noted: "The brain of a first-class genius like Gauss is as far removed from that of the savage Bushman as that of the latter is removed from the brain of the nearest related ape."[7] He recycled the quip in numerous articles and in almost every interview he gave, but it was not original. (He had in fact cribbed the line from Carl Vogt.) Nor did it follow from his own data, a fact that any casual reader can see in the drawings he carefully prepared. Spitzka himself would eventually notice that scientific racism was not as scientific as he first assumed, and he would find himself backsliding. After examining the Inuit brains acquired from the Museum of Natural History, he had to concede a new respect for that race, and somehow explain it in light of his former opinions. Without looking too hard, he managed to "discover" the natural intelligence and high artistic talent of the "Eskimo."

Like all anatomists who preceded him in this line of investigation, Spitzka failed to find the anatomical locus of genius. But this did not prevent him from claiming that he had. Although he admitted that it was "unwarranted to propose conclusions of wide significance upon so little material," he concluded that the anatomical substrate of high intellect resided in an interesting feature of the brain called the corpus callosum. Its size, he said, clearly reflects the caliber of the intellect.

The term *corpus callosum*, in its original sense, referred to the brain's white matter. In its modern sense, it refers only to the band of fiber bundles that connect the left and right hemispheres at a narrow passage in the middle of the brain. When a brain is cut down the middle, the callosum shows up in cross section as a white region roughly the shape and size of a section from a bell pepper. Yet its modest size, relative to the massive hemispheres, belies a significance on par with, say, the Panama Canal relative to the oceans it connects. It is the conduit that allows the brain to talk to itself.

The older sense of the term is more appropriate to Spitzka's theory, or, more accurately, to his father's theory. In the 1880s, Edward Charles Spitzka had ventured to suggest that the amount of gray matter in a brain is not as important as the amount of white matter. What distinguishes the brains of men from mice, he said, is not size, but the ratio of white over gray matter. "The greater the relative preponderance of alba (white matter) over cinerea (gray matter) the higher the intelligence," he wrote. Like all theories of mind and brain, this one relied on a metaphor, and the senior Spitzka provided it: "White matter," he declared, "means elaborated and individualized projection of gray matter, as a multitude of parallel telegraph wires means a multitude of stations. If the telegraph wires are in number out of proportion greater than the number of stations, it means a more intricate inter-connection of station to station."[8] Like all previous theories of the intellect, it was no more than a guess, but it was a guess that could be tested, and the younger Spitzka thought he knew how to do it.

"If in the brain of an average man," he wrote, "there be a hundred, or two hundred, or five hundred connections for every fact that he remembers, their number is many times greater in that of the intellectually superior genius." All he had to do was look for this elaboration of brain structure in the cross sectional area of the corpus callosum.

The theory got an immediate boost when Joseph Leidy's brain, the most important in the collection, turned out to have the largest callosal cross section of all. Moreover, Spitzka found that his eminent men had larger cross sections, on average, than his control group of criminals. Even though he was working with a small sample of brains, these findings made him bold enough to announce the discovery of "an index in somatic terms of how we may distinguish the brain of a genius or a talented man from that of a person of only ordinary abilities."[9]

Two days after announcing this conclusion, Spitzka came under attack. "I believe Dr. Spitzka has erred in assigning to the callosum the high use he has designated," said an anatomist who preferred to remain anonymous.[10] (He was quick to add, "I cheerfully ascribe, however, to the general estimate of Dr. Spitzka as a thorough scientist who has contributed in this book one of the most valuable works of research of the past decade.") Milton Greenman, the director of the Wistar Institute, cast further doubt when he noted that he had seen many instances in which normal people had brains with no corpus callosum at all. He praised Spitzka's efforts, but called his theory of the callosum "something that will bear further investigation," which was nothing more than a polite form of dismissal.

To Spitzka's chagrin, the only substantial "further investigation" of his work came about when scientific racists cited it as an argument for white superiority, a development he could not have anticipated. In the ensuing furor, Spitzka would be caught in the crossfire, and his magnum opus would be ripped apart.

IN 1906, AN obscure Virginia physician named Robert Bennett Bean published a study in the *American Journal of Anatomy* which claimed to prove the impossibility of educating African Americans. Bean's argument rested, at least in part, on Spitzka's theory of the callosum. To make matters worse, Bean found a much wider audience when he adapted his paper for publication in the popular *Century Magazine.*

None of this sat well with Franklin Paine Mall, Bean's former mentor at the Johns Hopkins University School of Medicine, and the cofounder of the journal in which Bean's study appeared. At the time, Mall was the preeminent embryologist in America. He had been brought to Johns Hopkins by William Osler, and had established the country's first department of embryology. He also sat on the board of trustees of the Wistar Institute, and served as president of the American Association of Anatomists. Which is to say that Franklin Mall's impeccable credentials gave him the clout to take on his former student. At the very least, he thought it his duty to rescue anatomical science from propagandists.

Bean's argument was based on a false correlation. "Due to a deficiency of gray matter and connecting fibers in the negro brain, especially in the frontal lobes," he wrote, "we are forced to conclude that it is useless to try to elevate the negro by education or otherwise."[11] Mall had seen enough brains to know that this wasn't true, but he was forced to mount a study of his own to prove it.

Few scientists of the time felt compelled to attack their colleagues' racial beliefs, but Mall was especially troubled because Bean's work was badly flawed, and his intentions were clearly reprehensible. Mall was also concerned that the public was naive enough to believe it. (But not the press, it seems. In 1911, the *New York Times*, in a juxtaposition that seems too telling to be accidental, ran this headline: "To Have Us Know More About Brains—Future Anatomists May Profoundly Affect Our Political System—Why the Negro Is Slow to Rise." It was placed next to this one: "Four Lynched in Georgia—Three Negroes Taken from One Jail—A Sheriff Knocked Down."[12]) Bean's work was not only bad science, but was bad *for* science, and especially for anatomy. Mall had worked too hard to bring legitimacy to his profession to allow it to be dragged down in this way. He had to set the record straight.

As Mall began his own investigation, he soon discovered a larger problem. The literature of comparative brain studies was riddled with methodological errors that had led some anatomists to claim, rather outrageously, that "they could determine, almost at a glance, whether or not a given specimen came from a great man, a woman, or from a negro." Mall decided that he would have to cast a wider net to address these misconceptions. He published his findings in 1909 in the *American Journal of Anatomy* under the title "On Several Anatomical Characters of the Human Brain."

One of the most common misconceptions of Mall's era concerned the size of the frontal lobes. Edward Spitzka had voiced the majority opinion when he asserted that the forebrain is more prominent (constitutes a higher percentage of total brain volume) in males than in females, in whites than in blacks, in eminent as opposed to ordinary men. Spitzka also subscribed to the significance of fissural complexity and the development of the convolutions. According to this theory, less-evolved people have brains that more closely resemble the smoother brains of apes, whereas the top of the evolutionary scale is represented by richly convoluted brains of the "Gauss type." (When confronted with the smooth, nearly unfissured brain of Chauncey Wright, Spitzka confidently proclaimed it to be the exception that proves the rule.)

Neither of these ideas is taken seriously today, and, after examining the evidence, Mall rejected them. In study after study, he saw conclusions contradicted by the very data that had been trotted out in support of them. Spitzka's theory of the corpus callosum was no exception.

"A few years ago," Mall wrote, "the startling announcement was made by Spitzka that the area of the cross section of the corpus callosum was larger in eminent men than ordinary men. Since the corpus callosum is associated mainly with the frontal lobe the observation, if correct, would be of great significance."[13] But Spitzka had failed to correct his measurements for overall brain weight, Mall noted. Nor had he tested a sufficient number of brains. His entire argument seemed to be based on one exceptional case—that of Joseph Leidy. "The rest of the callosa of notable men given by Spitzka are not above average for brains of the same weight," Mall continued. From Spitzka's own drawings, and from studies he conducted himself, Mall concluded that "many negroes of lighter brain weight have larger callosa than most of Spitzka's eminent men."

When he turned his attention to Robert Bean, Mall found his study to be more mean-spirited than Spitzka's, and every bit as porous. Bean had essentially combined Spitzka's "discovery" of the callosal-intelligence connection with the assertion that the frontal lobes contain the higher cognitive powers. The corpus callosum, it turns out, has an anterior part called the genu, and a posterior part called the splenium. Its overall area, Bean realized, is mostly a function of body size, and thus could have little to do with intelligence. What did matter, he decided, was the ratio of the area of the genu to that of the splenium. Bean believed that the frontal lobes were the key to rational thought. "Here probably reside the highest developed faculties of man," he wrote, "the motor speech center for the command of language; will power, the power of self-control, the power of inhibition and perseverance; the ethical and aesthetic faculties; and the power of thought in the abstract."[14] These faculties, he claimed, were the very ones in which the white race excelled. By contrast, "the negro is primarily affectionate, immensely emotional, then sensual, and under provocation, passionate"—the very qualities he associated with the posterior part of the brain.

Bean had measured the areas of the genu and splenium in a series of brains, assuming that they would reflect the relative strength of frontal to anterior functions in each individual. What he found was that the brains of white men did indeed have a larger ratio of genu to splenium as compared to those of blacks.

Mall was immediately suspicious. The data seemed too good, in fact so good that he decided to repeat Bean's study, but with an important difference. He assigned a number to each specimen—the very same specimens used by Bean—so that he would not know either the race or sex of the subject as he made his measurements. The resulting data did confirm one of Bean's propositions—that the size of the callosum increases with brain weight—but none of the others.

"My findings," Mall wrote "do not confirm Bean's result that the genu is relatively larger and the splenium relatively smaller in the white than in the negro brain."[15] Where Bean's data had shown, via a statistical chart, a complete separation of white and black brains (where the size of the genu was plotted against the size of the splenium), Mall's data showed a random scattering of both races, with no separation. Where Bean had found a perfect correlation, Mall found none. The same held true when comparing the sexes. The results remained consistent for every proposed anatomical marker: brain weight, callosum size or configuration, fissural complexity, robustness of convolutions. Mall was unable to detect the sex or race of an individual in the morphology of his or her brain.

In retrospect, Bean's work falls squarely into the tradition of Galton's *Hereditary Genius,* Lombroso's *Criminal Man,* and more recently, *The Bell Curve,* by Charles Murray and Richard Herrnstein. It relied on logical fallacies that were easily exposed. Mall probably knew that the scientific refutation of such works never achieves the same impact, or reaches as wide an audience, as the originals. He could only try to set the record straight, to present the facts, to trust that they would eventually emerge from the fog of misrepresentation.

As for elite brains, another obsession of the previous fifty years, Mall was equally circumspect when he noted that, "The hope has often been expressed that through the study of the brains of men of genius anatomical conditions would be found which may account for their eminence."[16] But he could find nothing in the literature to gainsay what Rudolf Wagner had concluded from the beginning—that higher intelligence may exist in individuals with smooth brains or richly convoluted ones, heavy brains or light ones. The best investigators (here he singled out Gustav Retzius of Stockholm) had failed to find markers not only for intelligence, however defined, but for any special talents or skills. "I may be permitted to add," he said, "that brains rich in gyri and sulci, of the Gauss type, are by no means rare among the American negro."

He could not resist taking one last swipe at Spitzka in a footnote. In the line drawings Spitzka had prepared for "A Study of the Brains," Mall singled out the page that had inspired the infamous quote regarding the brains of Gauss, a Bushman, and a gorilla. The facing page featured Spitzka's drawings of the brains of General Skobelev, the German anatomist Richard Altman, and the French statesman Leon Gambetta. Comparing the two sets of illustrations, Mall wryly commented, "It appears that Gambetta's brain resembles the gorilla more than it does that of Gauss."

WITHIN A FEW years of its publication, Spitzka's "A Study of the Brains" receded from memory, but the Whitman story would not go away. Whitman's executors still wanted an explanation, and, with some persistence, they got one. It came from Henry Donaldson.

The trajectory of Donaldson's professional life had been set in 1890, when he examined the brain of Laura Bridgman. At the time, he knew that his published report was undermined by a lack of knowledge about the stages of the brain's development. The growth of Bridgman's brain seemed to have been stunted, especially in the area of the Sylvian fissure and the optic nerves, but Donaldson had no way of placing these facts in the context of her childhood illness, or of saying what her brain's development might have been without it.

By 1892, the year of Whitman's death, Donaldson was conducting research in physiological psychology (the branch of psychology pioneered in America by his mentor, G. Stanley Hall) at the newly founded University of Chicago. In 1895, he produced his first major work, *The Growth of the Brain: A Study of the Nervous System in Relation to Education,* in which he examined, among many other things, the brain weights of eminent men; he concluded that "the differences in weight are not to be associated with the difference in intellectual capacity, until the possible influence of social conditions has been excluded." (As he would later discover, it could not be excluded.) By this time he had joined the American Anthropometric Society (both he and Hall were invited to join in 1891) and had developed close ties to the Philadelphia brain men. This connection would pay off in 1906, when Donaldson was hired by the Wistar Institute as director of research.

Isaac Wistar, still the chairman of the board of directors, continued to lobby for elite brain research up until his death in 1905. He assumed that Donaldson would place the brain collection at the top of his agenda, but Donaldson was no longer interested. A preserved brain, as he later explained, is but "a crude machine lacking power and controls." It is also at best a mere snapshot in time of the ever-evolving life of a mind. In order to do meaningful research on the brain, he decided, he would have to find a better specimen, one that would allow him to chart the changes in one brain over the course of a lifetime. He found it in the albino Norway rat.

Donaldson began breeding rats at the Wistar Institute for experimental purposes after his discovery that "the nervous system of the rat grows in the same manner as that of man—only some thirty times as fast."[17] Rats allowed Donaldson to study the growth of the brain without having to worry about acquiring and possibly wasting priceless human specimens. Once Donaldson cast his lot with laboratory animals, the brains of the AAS became historical curiosities. So it must have come as a surprise when Herbert Harned, the son of Whitman's lawyer and friend Thomas Harned, contacted Donaldson with a request. Could he shed any light on the fate of Whitman's brain? As it turned out, he could.

According to Harned, "Dr. Donaldson looked the matter up in the files of the Wister [sic] Institute, and told me that the records state quite definitely that the brain was accidentally broken to bits during the pickling process."[18] Unfortunately, the records in question no longer exist, quite possibly because they were not the records of the Wistar Institute itself, but of the Wistar-Horner *Museum.*

The Wistar Institute was never part of the University of Pennsylvania. It was built across the street from the university's medical school, and its founders worked out an arrangement whereby it would remain an independent entity. The Wistar-Horner Museum, on the other hand, was the anatomical museum of the medical school. According to the arrangement, the new institute would acquire the anatomical collection and display it properly. But it did not open until the fall of 1894, two and a half years after Whitman's death, and long after the pickling of his brain would have been complete. Without knowing it, Henry Donaldson cleared the institute of all blame in the loss of Whitman's brain. At the time, no one put the facts together, but it seems that the brain had been dropped at the Wistar-Horner Museum in the old medical college across the street, where Francis Dercum and Henry Cattell had worked. The mystery of Whitman's brain, it seems, is thus solved.

HENRY DONALDSON'S BRAIN now resides in the basement of the Wistar Institute along with the brain of William Osler. In 1905, after fifteen years at Johns Hopkins, Osler left Baltimore for Oxford, England, to assume the position of Regius Professor of Medicine, and to be knighted as Sir William Osler. During the Great War he served as physician in chief of a Canadian military hospital. His son Edward, an artillery officer, was killed in Belgium in 1917, and, like Kipling after the death of his own son, Osler never recovered from the loss. Two years later, worn out by the war and by grief, he died of congestive heart failure. Out of loyalty to his fellow brain scientists, he had stipulated in his will that his brain should be removed and delivered to the Wistar Institute, and his wish was honored.

Eight years later, in "A Study of the Brains of Three Scholars," Henry Donaldson analyzed the gross anatomy of the brains of his old boss and mentor G. Stanley Hall, of the zoologist Edward Sylvester Morse, and of Osler. He did not dissect the specimens, and he limited his comments to their gross morphology. It would be the last study ever written on brains belonging to the American Anthropometric Society.

Donaldson tempered expectations from the start by pointing out that as interesting as convolutional patterns may be, they "can hardly be used to explain mental traits and abilities as between persons of ordinary and superior intelligence."[19] Several highly competent investigators—he singled out Gustav Retzius and Franklin Mall—had come to the same conclusion in scientific papers going back to 1896. As for the three scholars, he lavished praise on the men themselves, but of their brains he would say only that they were "well nourished." He was speaking literally. Proper nourishment and good health, it turns out, go hand in hand with larger brains.

A small chunk of Osler's brain, part of the occipital lobe about the size of a lemon, now sits in a jar of preservative at the nearby Philadelphia College of Physicians. It was recovered after a series of unfortunate accidents that typify the plight of many anatomical preparations. The chunk had been removed as part of a histological study conducted by the Canadian neuroscientist Wilder Penfield in the 1950s. It later turned up at the home of a former curator of the Army Medical Museum, who over the years had stolen enough specimens to fill every room of his house.

HAD WHITMAN'S BRAIN survived, it would have been examined, described, drawn, and praised by Edward Spitzka, along with the six other brains of eminent scientists and scholars in the Wistar collection. It would then have been returned to the vault in the basement of the institute, and would probably have been forgotten.

By 1901, the two living founders of the AAS, Francis X. Dercum and Edward Charles Spitzka, could not tell the younger Spitzka where the brains were being kept. They had no idea. In 1908, when Edward Anthony tried to reconstruct the history of the society, he could find no archives, no records of any kind. As late as 1960, the Wistar Institute tried to give the brains away, but there were no takers. The founders of the Brain Society, it seems, had acted on the enthusiasm of a fleeting moment in the history of science. Ironically, had someone not dropped Whitman's brain, they would hardly be remembered at all.

As for Spitzka, he never recovered from his failure to solve the riddle of eminent brains. By all accounts he was a charming man, well spoken, a good friend, but not an easy one. He was tall, athletic, and self-assured, with little patience for small talk. He was also a workaholic who compensated for the forfeited pleasures of youth, home, family, friends, and holidays by drinking. It is not hard to guess why.

It is not clear whether young Edward Anthony chose his profession or whether it was chosen for him, but it is safe to say that he was groomed for it, and in taking up his father's research interests Spitzka forfeited any chance of a life of his own. He would marry and have a son, but most of his time would be spent in the classroom or the lab. The opportunity to teach anatomy at Jefferson Medical College, an unprecedented offer for so young a man (he had just turned thirty), would bring with it the jealousy of his colleagues, intense pressure to publish, and the responsibility of training young anatomists. By the time he was installed as head of the new Daniel Baugh Institute of Anatomy in 1911, he was the editor of *Gray's Anatomy*, a job he held through three successive editions. In addition to his research he lectured, wrote scientific papers, prepared his own illustrations (he was a talented draftsman), managed a budget, raised funds, attended countless meetings of scientific clubs and societies, gave extemporaneous speeches, drank many a toast, and, at the age of thirty-six, suffered a nervous breakdown.

Officials at Jefferson blamed it on a combination of overwork and an unusual occupational hazard: death threats. Spitzka was probably the only scientist in America whose research papers were read by convicts, ex-convicts, their families, and friends. To cite just one celebrated instance, in 1908 he attended the execution of Chester Gillette, whose murder trial had caused a national sensation, and had inspired Theodore Dreiser's novel *An American Tragedy.* Spitzka removed and preserved Gillette's brain, published his findings, and the newspapers duly reported them.[20] The criminal underworld knew precisely what Spitzka was doing at the prisons.

It is not clear whether the death threats were real or a figment of Spitzka's clouded imagination. By the standards of Francis Dercum's *Textbook of Nervous and Mental Diseases,* his behavior had all the earmarks of alcoholic paranoia, although in his own mind the threats were real enough that he began to carry a gun. This much *is* clear: Edward Anthony Spitzka's fantastic rise in his profession never exceeded expectations. It couldn't have. His preoccupation with brains and his commitment to his father's obsession took him further and further away from the mainstream of neuroanatomical research. He had committed himself to gross anatomy and had advanced so swiftly that by the age of thirty it was too late to change course.

During the fall of 1912, Spitzka uncharacteristically began to miss some of his lectures. His friends began to notice a change. He had become furtive, convinced he was being stalked. On what would be his last day at Jefferson, he rushed into a crowded lecture hall at the Baugh Institute, slammed the door, and announced, "I'm being followed!" To the astonishment of his students, Spitzka reached into his coat and drew out two revolvers. After scanning the room, searching for unfamiliar faces, he satisfied himself that he was in no imminent danger and laid the guns down on the lectern. He then proceeded to deliver a brilliant anatomy lecture, well over the heads of everyone in attendance.

The Jefferson trustees agreed to an extended leave of absence. The newspapers picked up the story and ran it in bold type: "Brain Expert in Nervous Breakdown."[21] Spitzka disappeared for a while to Europe on the eve of World War I, returned to Philadelphia six months later, but did not resume teaching.

His father, still a practicing neurologist in New York City, had been suffering from cancer of the jaw for several years. It seems not to have slowed him down. Over the previous decade, Edward Charles Spitzka had contributed several articles to the *New York Times,* including a piece on a worldwide conspiracy to assassinate all of the major heads of state. According to the elder Spitzka, a criminal mastermind of Moriarty-esque proportions was pulling the strings of a vast anarchist network. He would not reveal his sources, except to say that they were well-placed European detectives.[22] He made his final appearance in the *Times* in November of 1913 as the target of a highly publicized lawsuit alleging that he had alienated the affections of a patient, the young wife of a jeweler.[23] The day before he was to testify in court, Spitzka suffered a massive stroke. His son arrived the next day, in time to be at his father's bedside when he died. After the funeral, the younger Spitzka boarded the train to Philadelphia, with a new specimen carefully stowed in his luggage.[24]

Spitzka submitted his resignation to Jefferson shortly thereafter. His career, which all along seemed to be merely an extension of his father's, then took its last fateful turn. He moved to New York, into his father's apartment, and took over his father's private psychiatric practice. He brought his impressive collection of brains, including his father's, with him.

IN 1912, SEVERAL months before his breakdown, Edward Anthony Spitzka had visited the studio of Thomas Eakins to sit for a portrait. It was a good match. The painter specialized in capturing the likenesses of eminent Philadelphians; the anatomist specialized in capturing their brains. Eakins had been painting the city's leading scientists, often on his own initiative, since the late 1870s, when he unveiled his first great medical painting. The subject was Samuel Gross (another Jefferson professor) in his surgical clinic. In 1888, he painted his most famous portrait, that of Walt Whitman. Now sixty-eight years old and in failing health, Eakins could barely manipulate his brush. He managed to rough in the outlines of a full-length study showing Spitzka holding a plaster cast of a brain in his right hand. Eakins's wife painted in the brain, the only part of the portrait that was finished. The rest is ghostly, unrecognizable.[25]

In the 1930s, a gallery owner cut the canvas down to size to make it more marketable, discarding in the process the portion that contained the brain. As it looks today, in a basement storeroom of Washington's Hirshhorn Gallery, it suggests a man who was once important for reasons no one can remember.

In 1917, Spitzka enlisted in the Army Medical Corps, served at field hospitals at home and in Europe, and was discharged in 1919 at the rank of lieutenant colonel. In his only scientific work after his breakdown, he published his last case study of an elite brain in 1918. He then relocated his family and (presumably) his brain collection to the suburbs of Westchester County. He had not done any real scientific research in the eight years since he left Philadelphia, although he was still technically the prosector for the AAS. Out of a fraternal sense of duty, he had stipulated in his will that this own brain should go to the society, and that his body should go to the anatomy institute he had once directed. He gave the same address for both—the Daniel Baugh Institute, Eleventh and Clinton Streets, Philadelphia—indicating his wish to take his place among the cadavers he had once dissected. His old mentor, Francis Dercum, still taught there, although it was anyone's guess who would be the director when his own body was delivered to the morgue.

Such thoughts were far away on a September evening in 1922, when Spitzka excused himself from the dinner table complaining of a headache. No sooner had he gone upstairs than, without warning, a blood vessel burst in his head. He was only forty-six years old.

In response to a query about the American Anthropometric Society in June of 1920, Francis Dercum, still the society's president, had replied, "The brains, which I am glad to say did not accumulate very rapidly, were deposited in the Wistar Institute of Anatomy, and in

1907 six were studied by Dr. E. A. Spitzka."[26] The remark is typically vague. Did he mean that he was glad that the members did not die off in great numbers, or was he expressing relief that the venture hadn't amounted to much? Probably the latter. After Spitzka's death, Dercum did not step forward to claim brain or body. Nor did he attempt to retrieve the brains of society members still in Spitzka's possession. Spitzka was buried at Mount Vernon Cemetery in Westchester County, and his sizable brain collection, numbering over six hundred specimens, vanished.

In a final irony, in the same year, half a world away, Vladimir Ilyich Lenin, who was plagued by crippling headaches brought on by overwork, suffered the same indignity as Edward Anthony Spitzka, only worse—he survived. He would suffer two more strokes in the ensuing two years before succumbing to a convulsive attack. His body would be preserved, rather famously, but few people are aware of the fate that awaited his brain.

Chapter 13

Lenin

TUCKED AWAY ON a winding side street within walking distance of Red Square there stands a rather shabby five-story brick building that, like many older buildings in Moscow, offers the barest hint of the deep secrets it contains. It has seen better days. In the courtyard formed by its L-shaped footprint, an accidental forest of uncultivated trees press their roots into the foundation, infiltrating and weakening a high retaining wall to the extent that it seems poised to collapse onto the sidewalk. At the toe of the L the building steps off its perch to meet the street at an unassuming entryway identified only by the number 5 and a small dark plaque with bronze letters reading: INSTITUT MOZGA, which the Russians, with their indifference to articles, translate as "Institute of Brain." To render this more simply as "Brain Institute" would be a mistake, because the missing article is definite. When properly inserted—as in, the Institute of *the* Brain—it implies something not at all generic.

Originally, No. 5 Obukha Lane housed a hospital. Today it houses "the brain." Institut Mozga was in fact founded for the express purpose of explaining one very special *mozg*, the most important of the Soviet era. Its original quarters still stand across town, in an elegant palazzo that now houses the French embassy, and is one of Moscow's architectural treasures. The contrast with the present building could not be greater, and it serves to emphasize a reshuffling of priorities that took place in the 1940s and has never looked back. In its crumbling new home, the Institut Mozga now merely struggles to survive and to justify its existence as something other than a historical curiosity. Although its research agenda has since diversified and its founding purpose is merely a footnote in its current prospectus, "the brain" is still there, and its examination continues to this day, over seventy years after it was removed.

On the fourth floor of the institute, the hospital's old chapel has been converted into a small museum of anatomical specimens and memorials to past researchers, in effect a reliquary with homage to the saints. It also serves as a colloquium room where visiting scholars present their papers. In winter, the temperature fails to rise above a parsimonious fifty degrees. When the display cabinets bordering the room like so many stations of the cross are illuminated, one in particular, containing the improbably large brain of a blue whale, stands out from the rest. Floating in its massive jar, it bears an uncanny resemblance to a human brain enlarged two or three times. A child might innocently assume, based on size alone, that this is the brain the institute was created to house. But it is not. Instead the eponymous brain resides farther down the corridor behind a door bearing the number 19. Compared to the whale brain, it is something of a letdown, having been sliced into paper thin sections and mounted on a series of ten thousand glass slides that fill an impressive cabinet. In a nearby cupboard fifteen thick volumes bound in green leather and containing 750 microphotographs constitute a highly detailed atlas of this hallowed specimen, whose owner's identity is spelled out in bold Cyrillic characters on the spine of each book, just as it is engraved on equally impressive monuments scattered throughout the city. It is, of course, the brain of Vladimir Ilyich Ulyanov, better known as Lenin.

Room 19 is padlocked. For years a KGB agent was assigned to it and monitored every person who entered. For in addition to the brain of Lenin, Room 19 contains a literal brain trust of the Soviet era, including the sectioned hemispheres of Joseph Stalin, the poet Vladimir Mayakovsky, the writer Maxim Gorky, and the physiologist Ivan Pavlov (the man who discovered the conditioned reflex). In a closet are jars containing whole brains waiting to be sectioned. The labels read like a pantheon of Soviet achievement, and include the filmmaker Sergei Eisenstein, the theater director Konstantin Stanislavsky, the novelist Andrei Bely, and the aviation pioneer Konstantin Tsiolkovskii. The most recent addition to the collection is the brain of the physicist Andrei Sakharov, who lived just around the corner from the institute in an apartment building that now bears a modest plaque in his memory. His widow, Elena Bonner, gave her consent at her husband's deathbed, and his brain now sits in Room 19 awaiting sectioning.

In all, there are thirty-four so-called "elite" brains in Room 19, not counting that of Joseph Stalin, which, despite the countless hours spent sectioning, mounting, and staining it, has been deemed unworthy of scientific study. In Room 19, as in all of Russia, Joseph Stalin has been demoted from the ranks of genius.

IRINA BOGOLEPOVA, a stout woman in her mid-fifties wearing a white lab coat of uncertain vintage, divides her time between her office, Room 19, and the preparation room, where her assistants operate the microtome. The instrument is older than the institute itself. It is the very device that was used to slice up Lenin's paraffin-embedded hemispheres in 1925, its wide, razor-sharp blade having been passed, meat-slicer style, tens of thousands of times over the specimen, and at least a million times over the embedded brains of writers, lunatics, artists, musicians, composers, politicians, murderers, children, apes, dogs, cats, and ordinary citizens. The procedure remains unchanged. The operator works diligently at slicing a wax-embedded block to a thickness of a few microns, oblivious to whose brain it is and not really caring, as she carefully places each slice on a glass plate. Meanwhile Irina Bogolepova peers into her microscope at the arrangement of cells in the cortex of an elite brain. Her specialty is cytoarchitectonics—the mapping of the cerebral cortex based on minute differences of cell structure, density, composition, and staining quality.[1] While every brain can be broken down into more or less the same general pattern of structurally distinct regions, there are significant variations from one specimen to another. Consequently, despite what the textbooks continue to insist, each brain has its own map, its own unique pattern.

What Irina Bogolepova is trying to discover is whether brains of the elite type have a characteristic cytoarchitectonic organization all their own, and in particular whether their neurons age better than those of average brains. This refinement of the phrenological paradigm, taken to the cellular level, is difficult and painstaking work. To map a single brain takes over a year, and the mapping itself cannot be parceled out to a team. Its practitioners speak of cytoarchitectonics as more of an art than a science, and history bears this out. No two researchers working with the same brain material ever produce quite the same map, and the many attempts to produce a definitive map of the functionally distinct regions of the brain, of which there have been several over the last century, have resulted in anywhere from fifty to over two hundred such regions with highly contested borders, much like the map of Eastern Europe itself.

Institut Mozga is the only laboratory in the world where elite brains are still actively being probed for, as one Russian study has referred to it, a postmortem diagnosis of geniality. Even within the institute, this research is the province of only a few scientists, notably Irina Bogolepova and her brother, Nikolai Nikolayevich Bogolepov, the institute's director since 1995. Together they have analyzed the brains of Pavlov and Mayakovsky. Unlike their cohorts at the institute who pursue more mainstream areas of neurobiological investigation, and who publish their results in Western journals, the Bogolepovs have a limited choice of outlets for their work. Thus far their brain studies have appeared in obscure Russian journals that are unknown beyond the country's borders. The Bogolepovs insist that they are not yet ready to discuss their most recent work. They are reluctant to admit that they have reopened the investigation of Lenin. They do not wish to compromise their results by prematurely leaking their findings. And that is all they will say about the matter.

THE STRUCTURE OF gray matter—not its surface appearance but its cellular composition—first came to light in 1840, when a French physician named Jules Baillarger sliced some brain tissue by hand, slipped it between two glass plates, and held it up to the window. Without a microscope, Baillarger was able to distinguish the six cell layers of the cerebral cortex, as well as the interconnectivity of the white and gray matter. "At the summit of the convolutions," he wrote, "the white matter is entirely united to the gray matter by many fibers. A simple juxtaposition of these components is thus inadmissible."[2] Sixty years later, a Welsh anatomist named Grafton Elliot Smith, using a similar mounting technique and a handheld lens, was able to divide the human cerebral cortex into forty structurally distinct areas based on subtle variations in cell layering. This was the first cytoarchitectonic mapping of a brain.

But before cytoarchitectonics there was myeloarchitectonics. In the 1880s, the German psychiatrist Paul Flechsig employed a very different method to map the sensorimotor cortex. The name he chose combined the prefix "myelo"—referring to myelin, the fatty axon sheaths that make the brain's white matter white—with "architectonics," a term employed by several researchers, including Rudolf Wagner, to describe the minute structure of the brain.

Flechsig noticed that during the first months of life, myelin sheaths develop gradually, and in a programmed sequence, around the long, whiplike tails of brain cells. Suspecting that what he was seeing was the physical manifestation of cognitive development, a sign of the maturation of neurons, he charted the sequence of myelination, matched it up against behavioral changes in newborns, and came up with a map of localized cerebral functions. It was not a phrenological map. Flechsig could identify only those areas that clearly related to sensory or motor abilities, and not personality traits. By a process of elimination, he concluded that thinking must go on in the last regions to myelinate, and he designated these the "association areas." It was a classic illustration of experimental physiology—comparing anatomy and neonatal behavior in order to assign function.

But while it was useful as a tool for marking the onset of some brain functions, myeloarchitectonics could not assist the search for the substrate of intelligence. It was limited to immature brains, and revealed nothing about intellectual achievements. Presumably, within each of the association areas of the cortex, finer anatomical divisions supported phrenological faculties, but finding them would require a different method, one that focused on the cell bodies themselves. Unfortunately, these cell bodies are effectively invisible under magnification. Their very existence was not even suspected until a new staining technique came along and gave researchers a peek into a previously unimagined world.

In 1872, Camillo Golgi, an Italian anatomist working in primitive conditions in the remote Tuscan village of Abbiategrasso, found a method for preparing and staining nerve tissue in such a way that a selection of brain cells stood out in a deep purple hue that would have made Tiffany envious. Golgi's silver chromate stain acted like a photographic emulsion to bring forth an amazing wealth of detail. (To this day, no one quite understands how it works.) With Golgi's discovery, microscopists could observe the minute cellular structure of the brain, could see individual cells, and, with this seemingly innocuous discovery, a new theory burst onto the scene. Only not immediately. It took fifteen years for the rest of the world to notice Golgi's technique, but, once that happened, the neuron theory quickly took shape. It was a theory that had been waiting for the right tools, and as Golgi's stain gave way to others, a host of new terms took root in what had been an uncharted landscape—neuron, axon, dendrite, glial cell, ganglion, and a host of classifiers: Betz cells, granular cells, pyramidal cells, and the Golgi apparatus—all of which were discovered and named in the 1880s. The neuron theory tried to account for the transmission of impulses through this network, to explain how an electrochemical charge could bridge the gap between adjoining nuclei (giving rise to the term synapse). It also presented a vast territory in which to look for, among other things, the anatomical basis of genius.

While Golgi's method was useful for isolating individual neurons, the most versatile stains were produced by the Frankfurt anatomist Franz Nissl in the 1890s. There were immediate benefits. Nissl's research collaborator, Alois Alzheimer, was able to identify and describe plaque deposits in the third cortical layer of patients who had died after suffering progressive dementia. Nissl's stains also led to more sophisticated ways of mapping the brain's cellular structure.

The first cytoarchitectonic maps based on Nissl's methods were made separately by Grafton Elliot Smith in 1905, and by an Australian psychiatrist, Alfred Walter Campbell, in 1907. Smith and Campbell made valiant efforts, but their maps failed to catch on for the most superficial of reasons: the labels they assigned to the cortical regions were difficult and unmemorable. As it turned out, the pioneer of record in the cerebral cortex would not be the man who drew the first good map, but rather the man who labeled the regions of the cortex using numbers instead of letters. His name was Korbinian Brodmann, and his fifty-one numbered regions, the so-called Brodmann areas, remain the standard divisions of the cerebral cortex to this day.[3]

Brodmann's map came along at a crucial time. Since the 1880s, anatomists worldwide had been naming parts of the brain independently, resulting in over fifty thousand arcane terms with numerous redundancies. On their own initiative, German anatomists took up the task of standardizing the terminology in the 1890s, but the Latinate polynyms they chose were difficult and obscure. (Try locating the gyrus occipitotemporalis medialis, for example.) So when Brodmann produced his cytoarchitectonic map of the cortex in 1909, he solved a long-standing problem. The awkward names gave way to numbers, and the numbered regions gave rise to a new phrenological map of the cortex.

WITH THE APPEARANCE of the neuron theory, elite brain studies were left in an odd position. After Edward Spitzka's failure, the prospects for distinguishing superior brains by their gross anatomy had almost been exhausted, and the phrenological paradigm was quickly losing credibility. Brains were still making their way into the collections, but they contributed little or nothing to the discussion. Most of them were placed on a shelf and never examined. In Stockholm, Gustav Retzius made some of the most careful studies of elite brains ever done, but unlike Spitzka, he refused to speculate beyond the evidence. He found suggestions of patterns—large parietal lobes in mathematicians, for example—but nothing consistent enough to constitute a trend. It seemed that the only place left to look was the cellular structure itself.

To the layperson, it must have seemed encouraging when, in 1905, a group of eminent researchers founded the International Brain Commission, whose goal was to coordinate the efforts of the various brain research institutes around the world, and to have each one assume responsibility for a neuroanatomical or neurophysiological specialty.[4] The membership constituted a who's who of neurology, and included Gustav Retzius, Camillo Golgi, Wilhelm von Waldeyer, and Henry Donaldson. This new group of brain men spoke freely to the press about the hundreds of elite brains in their possession, and of their hopes for locating the source of intellectual greatness.

Even the Japanese got into the act. In 1913, Prime Minister Taro Katsura, a keen follower of research trends in the West, initiated a brain collection by becoming the first donor. He would eventually be joined by the novelists Soseki Natsume and Yasuko Miyake, the haiku poet Naito Meisetsu, and scores of others.

But what the Japanese did not know was that elite brains were absent from the Brain Commission's list of research priorities. They were good for publicity, but not much else. The press was eager to publish the results of autopsies of famous men, and researchers like Donaldson, Retzius, and the others were happy to talk about the celebrated brains on their shelves, but their publications on the subject were limited to "observations" rather than discoveries. However tantalizing the prospects may have seemed at the time, elite brain studies remained bogged down in the pre-paradigm stage that Thomas Kuhn referred to as "early fact gathering." Nothing of any use had so far come of it, nor did it seem likely to.

The Tokyo collection still exists, and is housed in a museum at the University of Tokyo Medical School. With over 120 brains, it is still going strong. (The most recent addition, the brain of Prime Minister Takeo Miki, was added in 1988. But be forewarned: the collection is off limits to just about everyone.) Recently, its curator expressed some dismay when it was pointed out that in the ninety years of its existence, the collection has led to just two publications, both from the 1930s, and both dealing with fissural patterns. Over the course of almost a century, few researchers offered to study the brains for a simple reason that has plagued the endeavor from the start—the brains could not be dissected.

In Japan the problem is partly one of respect for the dead, but also, as is the case in the West, a desire to preserve the brains as relics. In order to justify slicing up and effectively destroying a brain, a researcher would have to lay out a research agenda with specific goals. But before cytoarchitectonics, there was no such agenda, and scientists were prevented from cutting into irreplaceable specimens (with occasional exceptions, as in the case of Laura Bridgman).

In the first decade of the twentieth century, however, cytoarchitectonics changed everything. It provided a way for a special brain to be both dissected and respectfully preserved at the same time. It also led to a new frontier on which to search for the anatomy of genius. All that was missing was a test case—that is, a sufficiently important brain and a researcher bold enough to slice it to pieces.

IN MAY OF 1922, six weeks after appointing Joseph Stalin as general secretary of the Central Committee of the Communist Party, Vladimir Lenin suffered the first of three strokes that would progressively weaken him physically, mentally, and politically. Too late, he realized that he had ceded too much power, that unless he recovered completely he would never get it back. A second stroke in December forced him to remain at his dacha in Gorky. His doctors, with Stalin's encouragement, forbade him to write. Still, Lenin pressed on, dictating letters and speeches day after day in a futile attempt to forestall a disaster, but to no avail. A third stroke suffered on March 9, 1923, deprived him of speech and completely paralyzed his right side, indicating massive damage to the left hemisphere of his brain. By September he could walk again. He even revisited Moscow for the last time in October, but he might as well have been a ghost.

A few years earlier, an oddly similar event took place in Washington, D.C., with very different results. On October 2, 1919, President Woodrow Wilson suffered a massive disabling stroke. The White House, at the behest of Mrs. Wilson, attempted to cover it up and kept the other branches of government in the dark. There was too much at stake. The president's inner circle, not wanting to risk exposure but desperate for guidance, decided to bring in a specialist they could trust. The list of candidates was a short one, and at the top of it was Francis X. Dercum, the man who had removed Walt Whitman's brain.

At that time, Dercum was the country's leading authority on the nervous system. He was also, by default, the president of the moribund American Anthropometric Society. Mere months after the Whitman business, he had left Penn's medical school to become a clinical professor of nervous diseases at Jefferson Medical College in Philadelphia. By the time of World War I, he was the grand old man of American neurology, having inherited the mantle once worn by Silas Weir Mitchell. In the tradition of generations of Philadelphia gentlemen, Dercum was the very soul of discretion. After examining the president, the avuncular neurologist calmly proposed not just a treatment, but a solution to a national dilemma.

The stroke had been serious, but limited (as were Whitman's and Lenin's) to motor and sensory functions. Wilson's mind remained alert, but he would need time to recover, and, during that time, he would have to be spared any mental strain or anxiety. Taking Mrs. Wilson aside, Dercum told her that to cede the presidency at this time could induce damaging stress. He advised her to carry on, to keep it quiet, to stall for time until the president came around.

"Madam," he said, "it is a grave situation, but I think you can solve it. Have everything come to you; weigh the importance of each matter, and see if it is possible by consultations with the respective heads of the Departments to solve them without the guidance of your husband." Years later, as Edith Bolling Wilson recalled, Dercum insisted that this was the most prudent course of action.[5]

"Had he not better resign?" she asked.

"No," replied Dercum, "not if you feel equal to what I suggested. For Mr. Wilson to resign would have a bad effect on the country, and a serious effect on our patient." Dercum, in effect, was rendering a medical and a political diagnosis. "If he resigns, the greatest incentive to recovery is gone; and as his mind is clear as crystal he can still do more with even a maimed body than anyone else."

This is how Edith Wilson, unbeknownst to the Congress, the Supreme Court, and citizens of the United States, kept her incapacitated husband in office, and in doing so became the de facto chief executive of the country over the ensuing six months.

LENIN WOULD NOT be so fortunate. Time was working against him. Even worse, his inner circle brought in not one but two dozen specialists. From Germany they summoned Oswald Bumke, Otfried Förster, Max Nonne, and Adolf von Strümpell; from Sweden, Salomon Henschen and his son Folke; and from Poland, Oskar Minkowski, to supplement a cohort of less-talented Russian physicians. As individuals, the Europeans were the best that money could buy, the most accomplished men in the field, many of them former members of the Brain Commission (which had disbanded at the start of World War I). But as a group, through their collective reluctance to take responsibility, they probably did more harm than good. While Francis Dercum had saved Woodrow Wilson's presidency by singlehandedly making a decision and pushing it through, none of Lenin's physicians had the clout to do the same for him.

In hindsight, there was one man who might have had the gumption to pull it off, to play the Dercum role. For a brief time, Lenin's dream team of foreign specialists had contemplated inviting Oskar Vogt to join them, and then just as quickly dismissed it. Vogt, the director of a neuroscientific laboratory in Berlin, had the requisite personality to deal with an irascible patient, and the backbone to stand up to a ruthless tyrant (as he would later do with Hitler). But Vogt was not a team player. He was used to being in charge, and his success in establishing the world's leading brain research institute outside of academia had not won him any friends on the inside. Most of the German specialists despised him and made a point of excluding him from the International Brain Commission. Vogt would be kept out of it—for the moment, anyway.

Meanwhile, all the State's horses and all the State's hired consultants, far from putting Lenin back together again, only made the situation worse. The patient banned all doctors from his sight. Otfried Förster, his primary care physician, was forced to observe Lenin through keyholes or from behind a screen set up near his bed. There was not much he could do. Lenin refused to cooperate, refused to rest, refused to cede control.

He died on the evening of January 21, 1924, after a convulsive attack. During his long illness, he had been attended by no less than twenty-six doctors. Nine of them attended the autopsy. At the time, it was thought that he might have been poisoned, that Stalin, impatient with Lenin's extraordinary powers of recovery, had hastened his departure. Unfortunately, the autopsy was delayed for sixteen hours and was so bungled that it will never be known for sure. Stalin made a point of being there to watch, which explains the sixteen-hour delay, by which time the brain had begun to degrade. Still, upon removal, it was found to weigh a respectable 1,340 grams. The assembled doctors were appalled at the degree of calcification in the blood vessels. One of them later commented that the scalpel sounded like it was touching stone. A dozen causes of death could have been cited, the mystery being not so much what had killed him as what had kept him alive. In the end, they concluded that Lenin had died of arteriosclerosis. The brain was preserved and taken to the Lenin Institute.

Semen Aleksandrovich Sarkisov, who would later become the director of Moscow's Institute of the Brain, once remarked, in a materialist-dialectical vein, that "the human brain is not only a natural-biological phenomenon, but also the social-historical product of the development of matter."[6] This was the official language of Soviet science, a language developed by Lenin himself. To the physicians who performed the autopsy, the social-historical significance of

Lenin's brain was not on the table. They suspected the cause of death, they expected to find damage in his left hemisphere (although not to the extent they did), they had no idea what was to come, nor did they dare to suggest the possibility that Lenin had been poisoned. They also failed to investigate the question of neurosyphilis, another rumor that would circulate in perpetuity (although the possibility has since been judged unlikely). The decision to preserve the brain was, if anything, in keeping with standard practice, a matter of opportunity more than conspiratorial motive. Only later did the Politburo conspire to make sure that the brain was studied. Their goal was simple: to find a materialist explanation for Lenin's genius. Lenin's embalmed body, by contrast, was merely a monument, a solemn relic of purely symbolic value. If dialectical materialism was correct, Lenin's great brain was the physical manifestation of a new ideology. Centuries of Russian history had been compressed and focused into this unique point in space, into this organ from which the destiny of the Russian people would now pour forth, along with the triumph of communism. Even if it was not put on display, like the Declaration of Independence, its symbolic value was incalculable. Only one piece of this grand picture was missing. The Politburo needed to find the right man to study it.

FROM THE START, the Russians had to face some unpleasant facts. All previous attempts to conduct a postmortem diagnosis of genius had been equivocal. They would have to try something new. They were also painfully aware that if one of their own, a Russian, studied the brain, the results would lack credibility with the rest of the world. They would have to secure the services of an outsider, but not just any outsider. Whoever studied the brain had to possess impeccable credentials, energy, organizational skills, and a sympathetic political outlook. He would have to appreciate the importance of their undertaking. They needed someone who knew how to get results.

As the discussions proceeded, the plan grew in complexity. If they were to go to the trouble of hiring a leading expert, they might as well pay him to set up a brain research institute in Moscow, and to train a core group of specialists who would be able to take over the institute someday. The task of recruiting this man fell to Nikolai Semashko, one of the physicians who had looked in on Lenin during his final illness. Semashko, now the people's commissar of health, would have a free hand to spend as much money as it took.

Among the team of Lenin's doctors, most were too old for the job. Otfried Förster was at the end of a long career. So was the Swede Salomon Henschen. For simplicity's sake, they would need a German. The new Soviet state had recently signed the Treaty of Rapallo, which provided for the exchange of scientific ideas with Germany, the first and only olive branch to be extended to the vanquished after the armistice of World War I. But Lenin's German doctors had had their fill of Moscow and had gladly left, clutching their scandalously high fees. The problem seemed insoluble until Lazar Salomonovich Minor, a prominent Moscow neurologist and a friend of Semashko's, remembered Professor Oskar Vogt.

In 1922, Minor had attended a conference in Berlin at which Vogt had announced two breakthroughs. The first, his theory of pathoclisis, explained the selective vulnerability to disease of certain regions in the brain. At the time, Minor thought this might be of some use in Lenin's recovery. The other discovery held out a different sort of possibility. Using the Nissl method, Vogt had succeeded in dividing the cerebral cortex into over two hundred structurally distinct cytoarchitectonic areas. Moreover, he had applied the technique to the brains of educated men and women, confident that it would reveal the anatomical substrate of genius. The presentation had made a strong impression on Minor, who passed Vogt's name along to Semashko.

OSKAR VOGT (no relation to Carl Vogt) was born in 1870 (the same year as Lenin) in Husum, a small town in northern Germany. Inspired by his father, who was a Lutheran minister, Vogt went off to college with the intent of studying philosophy, but soon switched to medicine, and earned a degree from the University of Jena in 1893.[7] He then embarked on a series of research apprenticeships with the most eminent men of the day. In Zurich he studied hypnotism and brain anatomy with August Forel; at Leipzig he learned myelogenesis under Paul Flechsig; in Paris he studied clinical neurology with Broca's disciple, Jules Déjérine, after which he hung out a shingle as a psychiatrist. Like many medical men of the era, he used medicine as his entrée into scientific research, starting with hypnosis, at which he quickly became adept, to the point of editing a journal—the *Zeitschrift für Hypnotismus.*[8] Vogt's interests and abilities ranged across psychiatry, experimental psychology, genetics, neurology, and philosophy, and he could have become a major player in any of these fields. But he never wavered in his choice of the most difficult of these pursuits (from a monetary standpoint)—the neuroanatomy of the human brain.

Vogt was a fireplug of a man—short, stocky, and fired with intense intellectual energy. A goatee and a rapidly receding hairline lent him an air of maturity that, together with his expertise and his well-heeled contacts, assured him a swift rise in his profession. It did not hurt matters that he had a good rapport with the opposite sex. His clients, rich women with hysterias, aging men with neuroses, impressed by the young man's pedigree, his discretion, his bearing, and his connections, underwrote Vogt's excursions into brain anatomy, genetics, genius, and the quest for the perfectibility of mankind. In return, he listened to their troubles, prescribed sedatives, induced healing trances, and gave soothing advice.

Vogt's first brush with fame occurred in the summer of 1896, when he accepted an offer to serve as resident physician at a spa in

Alexanderbad. Although a low-paying position, it would connect Vogt to a highly suggestible and wealthy clientele, some of whom, he could only hope, might be inclined to retain his services privately. As it turned out, one client would suffice. Margarete Krupp, the wife of the German industrialist who supplied Germany with its arsenal through two world wars, took to Vogt from their first meeting. Like many wealthy families, the Krupps had their share of personal problems, and in the young doctor Margarete found a man of judgment and talent beyond his years. She would become his sponsor, and in the ensuing decades this alliance would free Vogt from having to pursue a university position. With the Krupps' financial backing he would establish an independent research institute in Berlin, and the Krupp name would shelter his enterprises from war and fascism. Margarete Krupp, it seems, was destined to become Oskar Vogt's guardian angel.

During the same fateful summer, Vogt made another important connection when he hired an assistant, a young man fresh out of medical school. Although the assistant had hopes of becoming a country doctor, his plan had been temporarily derailed by a bout of diphtheria, and the summer job was supposed to serve as a form of convalescence. But he was soon infected with higher ambitions. After a few months of listening to Vogt's grand schemes while calming the frayed nerves of the haute bourgeoisie, he abandoned his dream and set off for Frankfurt to learn brain anatomy from Alois Alzheimer. The young man was Korbinian Brodmann.

At the end of the summer Vogt returned to Paris to work in the clinic of Jules Déjérine. It was there that he made the third crucial connection of his career when he met Cécile Mugnier, a tall, worldly Frenchwoman who was writing a doctoral thesis on myelogenesis. They married in 1899. Cécile would become Vogt's lifelong research partner, and would supply the technical expertise and insight that her husband (who was more of an organizer and a visionary) lacked. With the grounding provided by Cécile, Vogt's career took off. The couple moved to Berlin, and with Krupp's assistance founded their

Neurobiological Laboratory. Vogt had not forgotten Brodmann. He offered him a job in 1901, and over the ensuing decade the gifted assistant produced his celebrated cytoarchitectonic brain map under Vogt's gaze.

Oskar Vogt could have ventured into at least half a dozen fields of research, but he decided to concentrate his efforts on finding the neurological basis for human behavior. He did not have much of a temperament for careful experimentation. His true calling was as an administrator and idea man, and he left the hands-on work to detail-oriented assistants like Brodmann, who staked out the field of cytoarchitectonics, and his wife, who did the same in myeloarchitectonics. Vogt himself oversaw the operation and coordinated many specialties—anatomy, physiology, genetics, histology, and microphotography—under one roof. From modest beginnings in 1901, he set out to build the greatest brain research institute in the world, and, over the next twenty years, through sheer willpower, he succeeded.

IN THE FALL of 1927, near the tenth anniversary of the October Revolution, Oskar Vogt stood before a group of officials who had come to the Pantheon of Brains in Moscow in order to hear the results of two years' work on the brain of Vladimir Lenin. It was the first of two presentations Vogt would give to his new patrons. Given the surroundings, they might have expected a strange performance, and they would not be disappointed.

When he first arrived in Moscow in 1925, Vogt had been given the royal treatment. His institute was installed in an exotic pavilion that would not have been out of place at a world's fair. (The switch to the moribund hospital at No. 5 Obukha Lane did not occur until the 1940s.) Designed in 1896 by the architect Nikolai Pozdeyev, the Maison Igmounov features a jumble of pitched and rounded roofs, turrets, hanging arches, and a wide range of ornamentation that defies categorization. Critics refer to it, almost unanimously, as fantastic or grotesque. When Igmounov himself made the trip from his dacha to see it for the first time, he was so horrified that he returned home and never set foot in the new house. At the time of the revolution, the Maison Igmounov was occupied by an American company. After the Bolsheviks appropriated it, they installed a plasma research lab in one half, and gave the other half to Vogt. It had not, of course, been built as a research institute, and Vogt and his staff did almost no research there. The training of the Russian technicians, the slicing and preparation of Lenin's brain, and all of the microscopic examination had taken place at Vogt's neurobiological laboratory in Berlin. The Moscow institute was essentially a showroom—thus the Pantheon of Brains.

In truth, Vogt had not given Lenin's brain much attention during the previous two years. The logistics were overwhelming. While he was training his Russian assistants, his own staff had transformed Lenin's sclerotic cerebrum into something out of a first-year calculus course—a sum of a seemingly infinite number of cross sections. (Some estimates place the number of slices as high as thirty-four thousand, an almost unimaginable number, and an impossible one given the thicknesses involved. Vogt was always purposely vague about numbers.) Of the thousands of slices that were mounted, only a fraction, perhaps no more than seven hundred, were stained and photographed. Vogt barely had time to glance at them prior to his presentation.

In later years Vogt was reluctant to talk about his Moscow venture. He knew that it had damaged his credibility. He had backed himself into a corner, he was expected to say something about Lenin's brain, to explain its uniqueness in layman's terms, and the best he could come up with was an anomaly in one of the cell layers of Lenin's cortex.

"The structure of the third layer of Lenin's brain," he said, "consisting of pyramid cells, deserves special attention, for these cells were found to be especially well developed. This in itself explains the outstanding nature of Lenin's intellectual life."[9] This seemed to strike the requisite balance between specificity and vagueness.

Pyramid cells are a broad class of so-called projection neurons that constitute about three-quarters of the neurons in the cerebral cortex. In the third layer, their role in memory retrieval and mental quickness was (and still is) inferred mostly from Alois Alzheimer's discovery of plaque deposits in cases of advanced senility. (Vogt would later point to the atrophy of such cells in the brains of idiot criminals.) Vogt suggested that the presence of large pyramid cells, along with evidence of higher-than-normal connectivity, could explain "the wide range and the multiplicity of ideas that developed in the brain of Lenin, his capacity for quickly getting his bearings when confronted with the most complex situations and problems, and his unusual powers of intuition."

To spell it out more clearly, Vogt reached for a musical analogy. He compared the process of thinking to the notes of a musical score. The average person, he said, thinks "with the slow tempo of an average piano selection." By contrast,

> Lenin's brain activity can be compared with a whole wave of sounds, closely interwoven, rapidly tumbling over one another, yet so combined as to produce a mighty harmony. All these aptitudes form, in their entirety, the basis of what is universally recognized as a type of genius. Thus the key to a materialistic view of Lenin's genius has been found.

It was precisely what the Russians wanted to hear. Now all Vogt had to do was to prove it.

But he didn't try. Instead, he spent the next two years in Berlin attending to other projects. In his absence, S. A. Sarkisov, Vogt's most adept Russian protégé, ran the Moscow institute. As he and his fellow researchers immersed themselves in cytoarchitectonic investigations and collected more brains, they began to grow increasingly skeptical of their mentor's techniques. To his credit, Vogt published their criticisms in his own *Journal of Psychology and Neurology*. He was unconcerned. Meanwhile, Commissar Semashko began to grow impatient with Vogt, and the two men had a falling out. Vogt was an uncooperative subcontractor. During his presentation to members of the Politburo, he had dared to suggest that Lenin had suffered from a hereditary neurological weakness. He was also spending too much time away from Moscow.

It is not clear when Oskar Vogt began to sour on the project. To run his own institute he needed money, and sometimes he had to compromise his better judgment in order to get it. In one of his rare comments on the Lenin business, he said, "Naturally, I had to stress strongly the possible connection of the large neurons in the third layer of the cortex with the associative capabilities of Lenin to satisfy the expectations of the attending Commissar Semashko. On his report to the Politburo depended the support for our institute."[10] Yet he steadfastly believed in the future of cytoarchitectonics and its potential for one day revealing the anatomical basis of genius. Unfortunately, that day had not yet arrived, and his biggest client was growing impatient.

Two years later, in 1929, on the twelfth anniversary of the revolution, Vogt again stood before the same group of distinguished academicians and party officials at the Maison Igmounov, in order to deliver a follow-up report. Because he did not have much to add to his remarks of two years earlier, he made sure to say it at great length.

His most notable achievement during the interim had been the acquisition of a control group. For his first presentation he had brought two brains from his own collection as comparisons—that of the 1909 Nobel laureate Wilhelm Ostwald, a German chemist, and another belonging to a murderer. Now, through the efforts of his Russian colleagues, he could point to thirteen specimens, including the brains of the Armenian composer Alexandar Spendiarov, the philosophical writer Alexander Bogdonov, and Lenin's friend A. D. Tsurupa, the commissar of food. Nine of the brains had already been embedded in paraffin. In time, if all went according to plan, each would be sliced up and undergo a complete cytoarchitectonic investigation. But there was a problem, and Vogt felt that he had to address it.

It seems that four years earlier, in 1925, a Viennese neuroanatomist named Constantin von Economo had published a massive cytoarchitectonic atlas of the brain, in which he advocated a new method for preparing and mounting slices of cortical tissue. Economo had singled out Vogt when he claimed that "for elite brains as unique and irreplaceable entities, the method of cutting into slices should not be done at all, under any conceivable circumstances, because it means the complete devaluation of the specimen."[11] What he meant was that longitudinal slices of paraffin-embedded hemispheres are unsuitable for microscopic study. If true, it would mean that Vogt had destroyed Lenin's brain. Whether out of pride or fear, Vogt spent the first fifteen minutes of his talk defending his own work through the simple expedient of ridiculing Economo and his "method of small blocks."

CONSTANTIN VON ECONOMO was Oskar Vogt's opposite in almost every way. He was born in Greece in 1876, had grown up in Trieste, then settled in Vienna, where he divided his life between two passions: aviation, which he first took up in 1907 as a balloonist; and medicine, which he decided to study after reading Lombroso's *Genius and Insanity* at the age of fourteen. In addition to his native Greek, he spoke Italian, German, French, and English fluently. (Vogt, who had lived in Paris and even married a Frenchwoman, never mastered French, or felt comfortable in any language but his own.) Economo's apprenticeship, like Vogt's and so many others', took him to Paris, where, in 1903, he studied neurology under Pierre Marie (who was also Cécile Vogt's mentor). Unlike Vogt, Economo was cosmopolitan, handsome, titled, and debonair. As the Vogts were establishing their neurobiological laboratory in Berlin, Economo was settling into a similar institute in Vienna run by Julius Wagner von Jauregg, a 1926 Nobel laureate. It was there, in 1923, that Economo made the discovery for which he is still remembered today—encephalitis lethargica, or the sleeping sickness.

The contrast in their personal styles carried over to their working habits in a confounding way. The dashing balloonist was attentive to details, while the plodding Vogt ran roughshod over them. Economo labored for ten years to produce his cytoarchitectonic atlas, probably subtracting an equal number of years from his own life in the process.[12] Vogt, on the other hand, with his grand schemes and plans, and a wife and daughter who compensated for his shortcomings as a technician, would live to age ninety. The field on which their conflict played itself out in the 1920s was a remote corner of neuroscience the two men had to themselves. Needless to say, they despised each other.

Economo's critique of Vogt boiled down to a question of geometry. He insisted that all cortical slices should be made perpendicular to the surface, which could only be done by first cutting the preserved hemispheres into small blocks, encasing each block in paraffin or celloidin, and then slicing them on the microtome. In Vogt's method of serial slicing of entire hemispheres, only the initial slices at the longitudinal fissure were perpendicular to the surface. From a purely geometrical point of view, Economo's method makes more sense. Slicing at varying angles to the perpendicular alters the thicknesses of the layers. But as Vogt pointed out, cutting the brain (an ovoid) into blocks (essentially cubes) entails some waste. Part of the specimen would have to be sacrificed, a fact that Vogt found unacceptable. Economo countered that Vogt's method lacked precision. In serial sectioning, a few slices are perpendicular to the surface, a few are exactly parallel, and the rest span the ninety-degree sweep in between. No waste, but how useful were the results?

Two hundred fifty years earlier, Nicolaus Steno had warned that *how* an anatomist sees the brain depends, quite literally, on how he slices it, and once again, the issue of the proper way to carve up a brain had become a point of contention. It would prove to be the Achilles' heel of the cytoarchitectonic method: a disagreement on uniform standards of specimen preparation exposed the inherent subjectivity of the method, and would invite comparisons to phrenology, along with accusations that elite brain studies were more art than science. To their credit, both Vogt and Economo made important discoveries in neuropathology, but these would be overshadowed by their claims about cell structure, intelligence, and talent, which reached the height of absurdity in Vogt's second speech to the Russians. It still reads like an intentional parody.

Vogt loved metaphors. He once compared cytoarchitectonic investigation to flying over a landscape and trying to make out towns and villages from a great height. "Only the talented investigator of architectonics can quickly spot characteristics such as unusual buildings by which to identify particular towns," he said. Vogt did not try to hide the subjective side of the endeavor. "The observer should also appreciate the aesthetic side of cytoarchitectonics," he said, "because it is a requirement for spending day after day at the microscope cheerfully—and for making new discoveries, great cheer is a necessity."[13]

In his attempts to please his Russian audience, Vogt reached for a phrenological metaphor when he said, "We have observed in Lenin's brain significantly large and particularly numerous pyramid cells in the third layer in the same way an athlete is characterized by a strongly developed musculature." The inference is obvious enough—great thinkers have overdeveloped pyramid cells in the third layer of the association cortex just as sprinters have overdeveloped quadriceps, hamstrings, and calf muscles. Vogt had a habit of stating guesses as facts, of piling one assertion on top of another into a fantastic confection, culminating in this case with a memorable conclusion: "For all these reasons, our brain anatomic results allow us to recognize Lenin as an *association athlete.*"[14] The italics are Vogt's, suggesting that he had underlined the word *Assoziationsathleten* in his notes, and delivered it with an extra spritz of saliva (the hallmark of his in-your-face speaking style).

Years later, the renowned Canadian neurophysiologist Wilder Penfield would suggest a simpler explanation for Vogt's finding. While visiting Vogt in 1928, Penfield had looked over some of the slides of Lenin's brain. The swollen cells of the third cortical layer looked familiar. When he learned that the autopsy had taken place sixteen hours after Lenin's death, it made more sense. Brains tend to swell as they decompose. The large cells, Penfield concluded, were merely an artifact of a compromised preparation process.[15]

BOTH ECONOMO AND VOGT were eugenicists. Like Galton and Spitzka, they believed in the perfectibility of mankind through artificial selection. Economo called the study of brains of the gifted "one of the most important problems we have."[16] In a 1927 paper entitled "How shall we study the brains of exceptional individuals?" Economo tried to establish standards for the examination of these precious specimens. On a few procedural points, of course, he and Vogt disagreed, yet both concurred that, as Economo put it, "an accurate description of the physical and psychic personality of the examined individual would have to be given." It was a depressing verdict because of its sheer impossibility. A psychological profile of Lenin or of any of the founders of the Soviet state was out of the question.

Vogt informed his Russian audience that "the brains we have so far collected are from those with personalities that have not been analyzed during their lives. So this has to be done by questioning people who knew the deceased. For this purpose we have designed a questionnaire."[17] This unofficially marked the end of the investigation. It was too delicate.

Vladimir Mayakovsky, for example, although considered the poet of the Revolution and a staunch admirer of Lenin, had committed suicide in 1930, partly due to the interference of a Soviet leadership that had fallen out of step with his ideals. Stalin would rehabilitate Mayakovsky, not as a discontented and lovesick man, but as "the best and most talented poet of our Soviet epoch."[18] Sergei Kirov, a member of the Politburo, had built a political base in Leningrad that rivaled Stalin's in Moscow. He was assassinated in 1934, marking the start of Stalin's great purge. (Nikita Khrushchev would later implicate Stalin in the plot). Kirov's and Mayakovsky's brains joined Lenin's at the institute and were sliced and mounted. But while the brains went under the microscope, the men themselves did not. A complete psychological profile would have limited their usefulness as official state heroes. Nor would Lenin's brain be compared with any other brains. In no respect could it be found wanting. Vogt was smart enough to see that he had done the job he had been hired to do. It was time to leave.

NINETEEN HUNDRED THIRTY-ONE would not be a good year for elite brain studies. First Economo dropped dead of a heart attack at the age of fifty-five, less than a year after taking over a new brain research center in Vienna. Then Francis X. Dercum, the last surviving founder of the American Anthropometric Society, the man who had removed Walt Whitman's brain and had saved Woodrow Wilson's presidency, suffered a fatal heart attack while opening a meeting of the American Philosophical Society. His own brain was not preserved. (Nineteen hundred thirty-one, by the way, was also the year *Frankenstein* was released.)

The year began well enough for Oskar Vogt. He left the Moscow institute in the hands of his protégés and returned to Berlin, to what he thought would be a bright and promising future. A consortium of donors, including the German government, had built him a new, expanded research facility in the Berlin suburb of Buch, making him the director of the largest brain institute in the world. But no elite brain publications would be forthcoming. Two years later, he would be ordered to fire all of his Jewish employees. When he refused, he came under suspicion himself.

Vogt's work in Moscow paved the way for accusations of procommunist sympathies and of harboring communist spies. He was harassed, but he held fast to his research program and rebuffed any effort to interfere with it. But eventually Vogt was forced out of his beloved institute. Despite his protestations of loyalty to the fatherland, and his insistence that in working for the Soviets he had merely been following the directives of the German government, he fell victim to political infighting. In 1935, he packed his microscopes and specimens and left Berlin. Once more he prevailed upon the Krupps to bail him out by funding a new institute that was built in Neustadt, in the relative calm of the Black Forest. It was there that Vogt weathered the war years.[19] In 1939, at the age of sixty-nine, he was drafted into the German army as a private and assigned to do administrative work at a local hospital. His calculated incapacity to salute every sergeant or corporal who walked by earned him a release after a few weeks.

In leaving behind his Moscow venture, Vogt would try to reclaim some legitimacy for his elite brain studies, for his gerrymandering of the cerebral cortex into over two hundred distinct and functionally independent blocs. The rise of Adolf Hitler interfered with his quest momentarily, but even the Nazis could not stop Vogt. He would outlive most of them, would continue to collect elite brains while perfecting his techniques. Vogt and his wife dedicated their lives to unraveling the secrets of the human brain, and to that end they amassed the world's largest collection of elite brains, as well as its largest collection of bumblebees (to test his genetic theories). But through it all, Vogt was unable to outrun the shadow of Lenin and the memory of the performance he was obliged to put on for the dignitaries gathered at the original Institute of the Brain in 1929. Twenty years, two world wars, and some 20 million deaths after Vogt's Moscow debut, the world's enthusiasm for elite brains and for the perfectibility of mankind had reached its limit. Vogt never published a scientific paper on any of his elite brain studies in Moscow. That task would fall to Sarkisov, who wrote such a report in the 1960s, by which time no one seemed to care.

As for Lenin's brain, if at first it seemed to be nothing more than an interesting anatomical specimen, it quickly accumulated considerable political importance. If dialectical materialism was right, the brain was nothing less than the physical manifestation of a new ideology. Even if it was not put on display, its symbolic value was incalculable. The Soviet leadership was not alone in this belief. In their attempts to take Moscow during World War II, the German high command drew up a top-priority plan to secure the brain and bring it back to Berlin. During the 1960s, a group of Soviet dissidents entertained the notion of stealing it in order to prove that the entire regime, along with the ensuing decimation of the population, had sprung from the corrupted imaginings of a syphilitic mind. At one time or another, it seems that just about everyone coveted Lenin's brain. It has since, of course, undergone a sharp devaluation.

BETWEEN 1930 AND 1936, any Soviet scientist, writer, or Party member who had the misfortune to die in Moscow had a good chance of having his or her brain sent to the Pantheon of Brains at the Institut Mozga. Permission was not an issue. Under communist rule all cadavers belonged to the state, and senior Party officials who died in those years had the honor of being buried in the shadow of Lenin's tomb at the base of the Kremlin wall. They also had the honor of contributing their brains. Among those who made the sacrifice were the composers Ivan Svortzkov-Stepanov and M. M. Ippolitova-Ivanovna; the biologist Ivan Michurin, considered the founder of Soviet genetics; the German communist Clara Zetkin; the Japanese Comintern official Sen Katayama; the writer and art historian A. V. Lunacharsky; the poet Eduard Bagritsky; and the opera singer Leonid Sobinov.

The number of brains and the secrecy surrounding them have inspired many myths about the Institut Mozga. Most of them focus on Stalin, on the notion that he created the institute and personally monitored its progress, that he envisioned a eugenic program that would create a master race. But he neither created the institute nor had anything to do with the preservation and study of Lenin's brain. He had other business to attend to. Once Lenin was out of the way, Stalin cared mostly about his predecessor's body. His own brain interested him only to the extent that it would one day show him to be an even greater genius than Lenin. He would receive only one report on Lenin's brain, in 1936, and that would constitute his entire involvement with the institute until his own brain entered the collection in 1954.

As for the so-called "race brains" and the notion of a master race, Sarkisov and his colleagues steered clear of any implication that some brains are better than others. They echoed the findings of Franklin Mall, and demonstrated that there is no brain anatomical basis for distinguishing among the races. This was, of course, a foregone conclusion. The Soviet Union was a multiethnic state, and Stalin, a Georgian, had no desire to differentiate among its peoples.

After Vogt's departure in 1930, Sarkisov moved quickly to diversify the institute's research agenda and to expand the specimen collection to include more than human brains, so that by the end of the decade he had amassed the largest collection of primate brains in the world. He then embarked on an ambitious plan to map the cerebral cortex of humans, of apes, of dogs, and the like. After 1936, the pretext of studying elite brains in the spirit of dialectical materialism assured the continuation of funding, but acquisitions slowed to a trickle.

As for secrecy, the institute's activities have been a matter of published record since its inception, and the existence of its brain collection was common knowledge.[20] Visiting researchers from the West, many of whom worked in parallel with Soviet researchers, were routinely shown the brains, and in some instances they were encouraged to study them. For a time, a few of the brains were kept on display with the whale brain in the institute's small museum. Today, their "top-secret" status clings to them in the same way that the residue of Soviet bureaucracy still hangs over Russia—through sheer inertia.

Nikolai Nikolayevich Bogolepov took over as the director of the institute in 1995. His predecessor, Oleg Adrianov, had examined Lenin's brain, and although he saw nothing particularly unusual about it, he willingly discussed it with reporters. He gave them access to the institute and suffered their mistakes, perhaps conceding that any publicity is good publicity (especially when one's only source of funding has suddenly ceased to exist). When asked about the brain, he sometimes praised it. "The study of Lenin's brain gave birth to the Soviet science of the brain, and therefore a way of constructing a picture of the man, an undeniably great man," he said.[21] On other occasions, he hedged: "In the anatomical structure of Lenin's brain," he once admitted, "there is nothing sensational." The verdict was never clear, but Adrianov was unequivocal in his assessment of Stalin. "Frankly, we are not very interested in Stalin's brain." After Adrianov's death, Nikolai Bogolepov closed Room 19 to all visitors. The succession of nosy reporters had become a nuisance.

Vogt's own brain collection, perhaps the largest collection of elite brains extant, now resides in Düsseldorf, Germany, at the C. and O. Vogt Institute for Brain Research at Heinrich Heine University. The research director, Karl Zilles, continues the tradition of in vitro cytoarchitectonic study, now linked, using modern brain-imaging techniques, to in vivo studies. He regards Vogt's collection as being mostly of historical interest. Without in vivo data (meaning without brain scans of the living patient), the specimens have a limited value for cytoarchitectonic purposes. Professor Zilles declines to reveal the names behind the brains in the historical collection, mostly because they have not yet been inventoried. But some of them are known. Vogt's brain is there of course. So is Solomon Henschen's, one of the doctors who treated Lenin, as are those of the neuroanatomist August Forel, the poet and dramatist Hermann Sudermann, the translator Emil Krebs, who spoke more than sixty languages, and the distinguished psychiatrist Robert Sommer. For what it is worth, Vogt's bumblebee and beetle collections, which number some seven hundred thousand specimens, can be found at the Zoological Museum of Amsterdam.

As for Vogt's legacy, in the wake of the Lenin business, his cytoarchitectonic investigations came under increasing fire. In the United States, researchers had trouble reproducing his results. Karl Lashley of Harvard University showed that two workers, trained in the same methods and given the same monkey brain, produced two different cytoarchitectonic maps. Percival Bailey and Gerhard von Bonin, in their book *The Isocortex of Man* (1951), accused the Vogts of seeing subtle boundaries in the cortex where others saw nothing. The efforts of Vogt's former Russian students, they said, "serve to make our knowledge of the cortex anatomically top-heavy." The cytoarchitects had investigated cortical structures so obsessively, they added, that "the intrinsic function of the cortex was hardly ever asked."

Bailey and von Bonin not only questioned the validity of elite brain studies, but disparaged all earlier attempts to correlate anatomy with performance, or to assert the superiority, on a brain anatomical basis, of one race over another. "The safest conclusion to be drawn appears, therefore, still to be that all human brains look essentially alike," they wrote. "A few significant differences crop up here and there, but they fail to lead to any consistent grouping, and one is therefore led to suspect the methods of observation."[22]

> For this reason we reject the excessive parcellation of the Vogt, Economo, and Filimonov schools as misleading and insignificant. As one reads through their prolix descriptions, and has the

> misfortune to remember what he has read, one is either repeatedly shocked by contradictions or suffers from what the French psychiatrists call *le phénomène du déjà vu*, description after description sounding merely like paraphrases of the preceding one.[23]

It was in this atmosphere of accusation and counteraccusation that a pathologist named Thomas Harvey, oblivious to the details but fascinated by the overall prospect, decided to remove and preserve the brain of Albert Einstein.

Chapter 14

Einstein

THE BRAIN OF Albert Einstein has acquired a notoriety out of all proportion to its value as an anatomical specimen. It is not part of any collection, it did not motivate a new theory of brain function, nor has anything of scientific value resulted from its study. Instead it has become exactly what Einstein most feared: a pop-culture icon.

The strange journey of Einstein's brain began on the evening of April 17, 1955, when the seventy-six-year-old physicist was admitted to Princeton Hospital complaining of chest pains. He died early the next morning of a burst aortic aneurysm. As in the cases of Carl Gauss and Walt Whitman, the issue of permission to perform an autopsy is clouded by subsequent testimony. Thomas Harvey, the pathologist on call that evening, would later say, "I just knew we had permission to do an autopsy, and I assumed that we were going to study the brain." As reporters soon discovered, Harvey did not have permission. Nor did he have a legal right to remove and keep the brain for himself. When the fact came to light a few days later, Harvey managed to solicit a reluctant and retroactive blessing from Einstein's son, Hans Albert, with the now-familiar stipulation that any investigation would be conducted solely in the interest of science, and that any results would be published in reputable scientific journals.

But Einstein's dignity had already been compromised. He had left behind specific instructions regarding his remains: cremate them, and scatter the ashes secretly in order to discourage idolaters. Yet not only did Harvey take the brain, he also removed the physicist's eyeballs and gave them to Henry Abrams, Einstein's eye doctor. They remain to this day in a safe deposit box in New York City, and are frequently rumored to be poised for the auction block.[1]

Within months of the autopsy, Harvey was dismissed from Princeton Hospital for refusing to surrender his precious specimen. His assurances may have satisfied Hans Albert Einstein, but Harvey's boss, the hospital's director, perhaps gauging his man somewhat better, was not impressed by the plan, and Harvey's tenure as a pathologist came to an end.

It should be emphasized that Thomas Harvey was not a brain specialist. His understanding of the brain did not extend beyond the postmortem diagnosis of disease, atrophy, or injury. Which is to say that he had neither the means nor the expertise to undertake the study he had proposed to Einstein's son. Although his accounts of the incident have varied considerably over the years, it seems that he removed the brain at the request of his mentor, Harry Zimmerman, who was Einstein's personal physician. Why he kept it will never be known for certain, but it can be inferred from comments made to various reporters that Harvey was inspired by Oskar Vogt's study of Lenin's brain, and he had the vague idea that cytoarchitectonics might shed some light on Einstein's case. A simpler and more appealing explanation is that he got caught up in the moment and was transfixed in the presence of greatness. What he quickly discovered was that he had bitten off more than he could chew.

After losing his job, Harvey took the brain to a Philadelphia hospital, where a technician sectioned it into over two hundred blocks and embedded the pieces in celloidin using a variation of the Economo method. Harvey gave some of the pieces to Harry Zimmerman, and placed the remainder in two formalin-filled jars, which he stored in the basement of his house in Princeton. Occasionally, he would try to interest a brain researcher in his quest, but most of the inquiries he fielded came from reporters. Whenever they asked what was being done, Harvey would confidently proclaim that he was just one year away from publishing his results. He would continue to give the same answer for the next forty years.

Harvey's marriage soon fell apart, and he left Princeton in search of work.[2] After his wife threatened to dispose of the brain, he returned to retrieve it and took it with him to the Midwest. For a time he worked as a medical supervisor in a biological testing lab in Wichita, Kansas, keeping the brain in a cider box stashed under a beer cooler. He moved again, to Weston, Missouri, and practiced medicine while trying to study the brain in his spare time, only to lose his medical license in 1988 after failing a three-day competency exam. He then relocated to Lawrence, Kansas, took an assembly-line job in a plastic-extrusion factory, moved into a second-floor apartment next to a gas station, and befriended a neighbor, the beat poet William Burroughs. The two men routinely met for drinks on Burroughs's front porch. Harvey would tell stories about the brain, about cutting off chunks to send to researchers around the world. Burroughs, in turn, would boast to visitors that he could have a piece of Einstein any time he wanted.

In the early 1990s, Harvey returned to Princeton, his wanderings not quite over. What had merely verged on the absurd in the early days crossed the line in 1997 when he embarked on a cross-country road trip with a freelance magazine writer named Michael Paterniti. Harvey wanted to meet Einstein's granddaughter in California. Paterniti eagerly signed on as a driver, and the two men set off from New Jersey with Einstein's brain in the trunk of Harvey's Buick Skylark. When he met the granddaughter, Harvey toyed with the idea of giving her the brain. He even left it at her house accidentally. But she didn't want it. In the end, the two men and the brain parted ways: Paterniti to seek fame and fortune with his story, Harvey to seek peace of mind at his girlfriend's house in Princeton, and the brain to end its days at the pathology lab where it had all started some forty years earlier.

In his book *Driving Mr. Albert*, Paterniti describes Harvey as an eccentric if not quixotic man with a booming voice and a odd habit of laughing at inappropriate moments. The only time Harvey seems to act responsibly is when he returns to Princeton and entrusts what remains of Einstein's brain to the pathologist who holds his old job at the hospital. Having disposed of the specimen that has determined the course of half of his life, he can at last go home.

During the first thirty years of this strange odyssey, the study of Einstein's brain went nowhere. To his credit, Thomas Harvey stuck to his promise. At any time, he could have sold the brain piecemeal or whole for a quick profit. Yet he never stopped trying to find researchers willing to study it. Not many were interested. Some dismissed the idea as nonsense, as starry-eyed lunacy. A few others agreed to have a look. Harry Zimmerman, who possessed about a sixth of the specimen, found nothing unusual, at least not in the brain itself. He was not so sure about his former colleague, and he began to deflect reporters' questions by claiming that Harvey was dead.

It was not until 1985 that Harvey finally caught a break. A Berkeley researcher read about the brain and its eccentric guardian, and she contacted Harvey to request a piece of it. She had an intriguing idea.

DURING THE 1970S, a UCLA neuroscientist named Marian Diamond conducted a series of experiments involving cell counts in the brains of rats. She segregated the rats into two groups and placed one group in a mentally stimulating environment, while consigning the other to a deprived environment. Over time, she found that the enriched environment produced more robust brains, while the deprived environment literally starved the brains. In other words,

Diamond studied brain plasticity—changes in anatomical structure brought about by environmental factors. She measured these changes by painstakingly counting individual brain cells on carefully prepared microscope slides.

In the early 1980s, Diamond heard about the existence of Einstein's brain, and she wondered if a similar study might demonstrate the same phenomenon in humans. She proposed the idea to Thomas Harvey, who at first refused, then relented, and finally mailed off four numbered blocks from Einstein's cortex. These became the basis for an unusual comparative study.

Diamond published her findings under the title "On the Brain of a Scientist: Albert Einstein." The article appeared in an obscure journal called *Experimental Neurology* in 1985, thirty years after Einstein's death, and not a moment too soon for Thomas Harvey. It began with a startlingly unscientific claim: "Albert Einstein is generally conceded to have had one of the greatest scientific minds that ever existed." While not a particularly controversial statement, it is an unprovable one, and it unintentionally set the tone for what followed.

Diamond had assembled a control group consisting of eleven brains of former veterans hospital patients, all men, and all dead from non-neurological disorders. Their average age was sixty-four. She had requested from Harvey sample blocks from Brodmann areas 9 and 39 in the left and right hemispheres of Einstein's brain in order to "follow the lead provided by [Einstein's] introspection." The blocks came from the association cortex, "the last domains of the cortex to myelinate, indicating their comparatively late development," as Diamond put it.

Diamond would have preferred to use Camillo Golgi's staining method, the most revealing of the available techniques. But because the specimens had long ago been encased in celloidin, which undermines the Golgi process, she settled for a different, less-revealing stain. Her plan was to count individual neurons and two types of glial cells known as astrocytes and oligodendrocytes, to come up with four total counts: neurons, astrocytes, oligodendrocytes, and combined glial cells. From these counts she constructed three ratios comparing neurons per unit area to each of the three glial cell counts. She then compared Einstein's cell counts and ratios to the counts and ratios for the control group. Seven measures (four counts and three ratios) from four blocks resulted in twenty-eight possible statistical comparisons.

After checking her results, Diamond rejected all but the neuron-to-glial-cell ratio ("it was necessary to pool all glial cells counted to attain statistically significant differences," she wrote), reducing the twenty-eight possible tests to four. Of these, Einstein differed significantly from the control group in only one Brodmann area—number 39 of the left hemisphere—in which he had a markedly lower neuron-to-glial-cell ratio.

Glial cells are support cells for neurons. They continue to divide throughout one's lifetime, whereas neurons do not (or at least, in light of very recent evidence, not very much). Consequently, there are only two ways for the neuron-to-glial-cell ratio to increase. Either the neurons die off too rapidly (which happens to victims of senile dementia), or the glial cells increase in number. In her previous studies, Diamond had found that enriched environments result in a proliferation of glial cells, thus lowering the neuron-to-glial-cell ratios in rats. The same ratios in Einstein's brain, she writes, "suggests a response by glial cells to greater neuronal metabolic need."

It wasn't much. A significant result in one out of twenty-eight comparison tests was not enough to allow Diamond to cry "Eureka." But it was enough for her to say that she had found something that "might reflect the enhanced use of this tissue in the expression of [Einstein's] unusual conceptual powers." Of course, once the press got hold of the story, she might as well have cried "Eureka."

Almost anything connected to Albert Einstein is news, and Marian Diamond's article was no exception. It caused a sensation. The wire services picked it up and flooded the once-obscure researcher with phone calls. She became a celebrity. Experts and nonexperts weighed in with comments on her study. Promoted from conjecture to fact without the usual review process, it found its way into the latest editions of many introductory psychology textbooks. At last, a possible explanation for Einstein's genius! But the excitement was premature. It seems that not only was Diamond's work riddled with serious flaws, but so, quite possibly, was Einstein's brain.

AS EINSTEIN'S BIOGRAPHERS have claimed (and as Einstein himself confirmed), he began speaking late, sometime after age three, and remained inarticulate and possibly dyslectic through his early school years. From a neurological standpoint, this could have resulted from the late myelination of key areas of the brain involved in speech, including Brodmann area 39. According to a 1992 study by S. S. Kantha of the Osaka BioScience Institute in Japan, Diamond's cell counts suggest "a strong possibility of some kind of lesion in this specific speech-related area in Einstein's brain which could have resulted in childhood dyslexia."[3] In other words, instead of possessing a superior brain, Einstein may have started out with a severely compromised one that subsequently healed. The possibility is especially intriguing because it suggests a theoretical diagnosis that, had Einstein's parents been aware of it, might have led them to steer their son toward the civil service instead of academia. But this would have to remain a mere speculation because Diamond's study, according to Kantha, was fatally flawed. He faulted her control group (who were these people?), the specimen itself (what had it gone through during the previous thirty years?), and Diamond's sketchy presentation of the data (why ratios and not actual cell counts?).

In another peer review, Terence Hines of Pace University, writing in *Experimental Neurology* in 1998, echoed Kantha's criticisms. Diamond's study, he wrote, "is so seriously flawed that its conclusions should not be accepted."[4] Hines's complaint is the same one lodged against Paul Broca by Stephen Jay Gould in *The Mismeasure of Man.* If you look long enough, if you measure a sufficient number of attributes, and if you are highly selective, you can eventually find statistical evidence to support or defeat any claim.

Hines searched through Diamond's statistical analysis for signs of bias. They were not hard to spot. Of the twenty-eight comparison tests she could have carried out, Diamond reported only four, of which just one produced the desired payoff. "It would be surprising indeed," Hines noted, "if from a total of 28 different tests one did not obtain at least one 'significant' result" (especially when using a 95 percent confidence level). Diamond admitted that she rejected those tests that didn't suit her purpose, that failed to show a statistically significant difference between Einstein and the controls. This was the fine print that her readers ignored in their rush to judgment. Hines also faulted the control group. An anonymous and seemingly random group of veterans leaves much to be desired. Who were they? What did they die of? It didn't end there. Diamond also came under attack for irregularities in her preparation technique. Some critics questioned the thicknesses of the tissue slices and the effects this might have had on cell counts.

In light of these faults, only one conclusion emerges as incontestable: that without Einstein's name on it, Diamond's paper would have gone nowhere. Not that any paper that gives marquee status to a genius will automatically generate headlines.

In 1996, Britt Anderson of the University of Alabama waded into the fray in *Neuroscience Letters* with "Alterations in Cortical Thickness and Neuronal Density in the Frontal Cortex of Albert Einstein."[5] The title tells all. The part of the frontal cortex in question is Brodmann area 9, a block adjacent to the one that Harvey had sent to Marian Diamond. Anderson measured the thickness of Einstein's isocortex (the outer six layers of gray matter), and counted the number of individual cells in a unit square area in order to calculate the size and density of the cells. As a control group, he used the brains of five men, average age sixty-eight. (Einstein died at the age of seventy-six.) The result? No significant differences in either the cell counts or in the size of cell bodies. Anderson did find that Einstein's cortex was thinner, but this was offset by a tighter packing of neurons. If Einstein had had a particularly large brain, then perhaps the dense packing of neurons would indicate a higher-than-normal number of neurons in total. But he did not. By any measure, Einstein had a smallish brain. Most commentators place it "within the average range," but at 1,230 grams, it weighed almost the same as the brains of Whitman and Gambetta, at the bottom of the list of famous men.

Anderson scoured the literature to see if he could find anything that might vindicate this result, and he came up with this: Two years earlier a Canadian researcher had tried to account for the 15 percent advantage in brain size in men over women, given that there is no appreciable difference in IQ. Her explanation? That a more tightly packed cortex might provide an advantage in processing time. The tighter packing of neurons within function-specific modules might actually reduce the interaction time between brain cells, she suggested. In other words, reduction in brain size could result in economy of mental processing. (Suddenly, smaller brains were better.)

Thomas Harvey liked the idea enough to make another one of his offers. Maybe this Canadian researcher could deliver the good news that would allow him to return home at last.

SANDRA WITELSON, a psychologist turned brain researcher, works at McMaster University in Hamilton, Ontario, where she maintains the Witelson Normal Brain Collection. Since 1977, she has been acquiring normal, healthy brains by removing, preparing, and storing the specimens under identical conditions. Each brain is accompanied by a detailed psychological profile of the donor, and each donor is tested to make sure that his or her IQ falls within a range that qualifies as normal. By the late 1990s, Witelson had collected thirty-five male and fifty-six female brains. With them, she conducted studies of brain anatomical differences in men and women, in homosexuals and heterosexuals, in left-handers and right-handers, dyslectics and nondyslectics. In her laboratory, Constantin von Economo's quest for uniformity seems to have come to fruition. Sandra Witelson had assembled the perfect control group.

With such a collection at her disposal, Witelson could avoid the biggest pitfall that had undermined Diamond's and Anderson's studies. There could be no complaints about the control group this time. Harvey made his usual offer, one she couldn't refuse, and mailed off a celloidin-embedded block from Einstein's temporal lobe. Witelson's associate, Debra Kigar, devised an experiment similar to Britt Anderson's—a cell count—with a similar, disappointing result: no significant differences.[6] It seemed like another dead end.

But Harvey had something else to offer. In addition to the brain itself, or what was left of it, he had taken a series of calibrated photographs of the whole brain at the time of the autopsy. No one with a trained eye had ever examined them. Would she be interested? It wasn't much, but it was at least worth a look. So Witelson put away her microscope, spread the photographs out on the table, and saw it almost immediately. Something in Einstein's brain was very, very different.

In the summer of 1999, the prestigious medical journal *The Lancet* published Witelson's findings under the tantalizing title "The Extraordinary Brain of Albert Einstein." In Harvey's photographs, Witelson claimed to have found a possible explanation for Einstein's remarkable powers of insight in the fissural patterns of his brain. She had noticed an aberration when she tried to trace the Sylvian fissure, the major lateral crevice, up to its source. Just where she expected it to zig, it zagged. When a river does this, it can instantly transfer land from one owner to another, from one county to another. It turns out that something similar had occurred in Einstein's brain. Instead of running its usual course, below and ending somewhere behind the fissure of Rolando (which runs right down the middle of the brain, almost temple to temple), Einstein's Sylvian fissure flowed into Rolando's fissure, and effectively ended there. As a result, he did not have a parietal operculum, the island that usually forms between the two major fissures.

Witelson had never seen anything like it. She searched through her control group, through textbooks and case histories, and found nothing remotely comparable. Einstein's parietal regions were so different that she decided to measure them, to compare Einstein to her control group. Sure enough, Einstein stood apart. His brain demonstrated what Edward Spitzka would have called a redundancy of the inferior parietal lobe. In other words, he had more brain matter of a crucial type. Could this idiosyncrasy be the key that would unlock the mind that changed history and lay bare the secrets of the human brain itself? It had to mean something.

Or not. The editor of *The Lancet* tempered his enthusiasm by filing the piece under "Department of Medical History," as opposed to "Articles," "Research Letters," "Case Reports," or even "News." His caution was justified because "The Exceptional Brain of Albert Einstein" did not review a defunct branch of medical science as much as resurrect it.

TO LOOK AT a brain, at the fissured surface of the cerebral cortex, is to see an ingenious solution to a geometric problem. This was first noticed in 1845 by Jules François Baillarger, the man who discovered the six cell layers of the cortex. When Baillarger looked at brains of different sizes and contemplated their cortical areas, he recalled the fact that as a sphere increases in size, the ratio of its surface area to its volume decreases. This is because the formula for surface area involves the square of the radius, while the volume formula involves the cube of the radius. Because the cube of a number increases much faster than its square, volume increases much faster than surface area. What this means for brains is that if the cerebra of all creatures, great and small, were smooth, smaller animals would have, relative to their size, much more gray matter than larger animals. By way of compensation, as primates evolved (and as their brains grew), their gray matter kept pace with their expanding brains by developing fissures in order to pack more surface area into their skulls. The fissures formed, moreover, in a recognizable pattern.

In 1854, when Louis-Pierre Gratiolet used the principal fissures and a few minor ones to divide the cerebrum into five lobes, he understood that the division was somewhat arbitrary, a convenience based only partly on deep structures—the Sylvian, central, and parieto-occipital fissures—and partly on superficial sulci and their extensions. In all but a few unusual brains, the network of fissures and sulci do not meet up to form closed boundaries around the lobes. In most brains, for example, the fissure of Rolando (or central sulcus) does not connect with the Sylvian fissure, yet it serves as the boundary between the frontal and parietal lobes. In some places the boundaries of lobes are extensions of fissural paths, mere dotted lines. Consequently, the frontal, parietal, temporal, insular, and occipital lobes are not distinct organs. They are also, by the same token, not functionally distinct. As ingrained as the terms might now be, Gratiolet's lobes are arbitrary divisions named for the parts of the skull they abut, and are based partly on structure, partly on convenience, and not at all on function.

When Korbinian Brodmann drew his famous 1909 map of the cerebral cortex, he delved below these surface patterns into the cellular structure and found fifty-one distinct areas in each hemisphere. In some cases, the boundaries of these zones coincide with fissures, but in most cases they do not. The brain's surface, it turns out, paints a misleading picture of its deeper structure. Later attempts to replicate Brodmann's map produced altogether different maps. The Russians, for example, showed that cytoarchitectonic areas could vary greatly from brain to brain, so that Brodmann's map could not serve as a reliable guide to the structure of every brain. Instead, it merely provided a convenient way to specify localized parts of the hemispheres. Given this disparity between superficial and deep structure, how much stock should be placed in Witelson's conjecture? In other words, do the paths of the fissures mean anything?

This was the million-dollar question. At least it turned out to be worth a million dollars. And Witelson had a million-dollar answer. She explained the Sylvian detour as the possible result of early and rapid development of Einstein's parietal lobes. The inferior parietal lobe, she noted, "is a secondary association area that provides for cross-modal associations among visual, somesthetic, and auditory stimuli." In slightly more concrete terms: "visuospatial cognition, mathematical thought, and imagery of movement are strongly dependent on this region."

Although Einstein's wayward fissure appears to be rare, overdevelopment of the parietal area is not. Witelson noted that it also showed up in the brains of Gauss and the Swedish physicist Per Adam Siljestrom. Had she read deeper, she would have found references to similar overdevelopment in the parietal regions of Gauss's successor Peter Lejeune Dirichlet, and in Gustav Retzius's descriptions of the brains of the mathematician Sonya Kovalevsky and the Swedish astronomer Hugo Gyldén.[7] This is not to say that overdevelopment of the parietal region is a constant feature of mathematicians' brains. In fact, Einstein's fissural pattern is more consistent with the criminal brains studied by Moritz Benedikt than with the brains of scientists like Gauss. This pattern, referred to as a "confluence of fissures," was for a time thought to be the hallmark of a criminal mind. Although Benedikt's thesis was discredited, morphologists of the late 1800s continued to investigate and tabulate instances in which the fissure of Rolando ran into the Sylvian fissure. In these studies a number of other anomalies were tabulated: interruption of the Sylvian fissure, doubling of the Sylvian fissure, and doubling of the central

(Rolando) fissure. Many such cases turned up routinely in studies of criminals, of eminent men, and of course, of ordinary people. They proved to be so common (in the 2 to 5 percent range in the general population) that all attempts to correlate them with behaviors or talents failed in large group studies.[8] Given this, any attempt to draw a conclusion from the fissural pattern of a single brain is pointless. It is as easy to claim that Einstein had an exceptionally compromised brain as it is to claim he had an exceptionally well-organized one. Instead of being a born genius, he might simply have been a born compensator. He might even, if such studies are to be believed, have been a born criminal.

"This report clearly does not resolve the long-standing issue of the neuroanatomical substrate of intelligence," Witelson concluded. "However, the findings do suggest that variation in specific cognitive functions may be associated with the structure of brain regions mediating those functions." Of the two claims, the first one is by far the more compelling.

In the dearth of responses that passed for peer review of Witelson's study, no one bothered to point out that this was not a study of a brain, but of *photographs* of a brain. Her control group consisted, presumably, of actual brains (unless she used photographs of her control brains as well). How did she calculate the areas of the parietal lobes? Even more important, how did she determine the boundaries of the parietal lobe? Were these true cytoarchitectonic boundaries, or merely surface markings? Witelson did not address these questions, but it didn't really matter. Without a complete cytoarchitectonic investigation, there is no way to tell whether the surface pattern represented a true delineation of the regions in question. In the absence of a brain scan while Einstein was alive, it is impossible to pinpoint the cognitive functions that were strongly represented in his parietal lobes.

Then again, a picture is worth a thousand words, and sometimes a million dollars. The payoff arrived shortly after the article's publication in the form of a million-dollar endowment from an appreciative donor (an Einstein enthusiast), earmarked for a chair in neuroscience at McMaster. Not surprisingly, the chair was awarded to Professor Witelson.

IN ADDITION TO teaching cognitive neuroscience, Cornell University psychology professor Barbara Finlay is the curator of the Wilder Brain Collection. Admittedly, there is not much left to curate, aside from the eight brains she has placed on display outside the offices of the psychology department. Wilder himself never used the collection to make conjectures about genius or special talent. Neither did his successor, James Papez (a groundbreaking neuroanatomist who in the 1930s proposed a circuit of brain centers that are responsible for producing emotions). And neither has Finlay. She is happy enough to have saved Wilder's brain twenty years ago when Cornell was on the verge of tossing out the whole collection. But other brain researchers have not been so cautious.

After several decades of poring over the neuroscientific literature, Professor Finlay is convinced that anyone willing to look hard enough can find a scientific study that supports almost any conceivable claim about the human brain. It works like this: A study of five specimens might reveal a marker for, say, dyslexia. This is to be expected. Small samples tend to return a high frequency of positive results. That's the beauty of a small sample—plenty of room for error. Another paper then comes along to refute it, or the author simply abandons the claim. Of the tens of thousands of research papers published each year, some will trigger alarms in editorial rooms in much the same way that computer programs trigger buy or sell orders on the stock market floor. If an article features the key words "Einstein's brain," the alarms will sound, and the paper will be plucked from the pile. Few people will read it, but many will read *about* it, usually in distilled form, with any weaknesses filtered out, so that the claim, which is, after all, merely a conjecture, reads like an incontestable fact. Not that such papers constitute fraud. But they are routinely taken far more seriously than they deserve to be. The researchers themselves know this, which is why they feature Einstein's name prominently in the titles.

Any scientific study that begins by informing the reader that Einstein is "our most renowned genius" (Anderson), or that he "is generally considered to have had one of the greatest scientific minds that ever existed" (Diamond), or that he is "one of the intellectual giants of recorded history" (Witelson), renounces any claim to objectivity at the start. Its motivation is transparent. Still, the author is obliged to lay out the psychological framework to support these statements. Just what is a genius, or a great mind, or an intellectual giant? Without a working definition, the neurological evidence would have no context.

Anderson, for example, begins his paper by admitting that the relevant quality of mind—intelligence—"is not a mystical psychological process which confers smartness in proportion to its presence."[9] Instead, intelligence is a "psychometric index for the effects of neurobiological features which affect the efficiency of neural operations." Such features exist, Anderson claims, because correlations between intelligence and the brain's anatomy and physiology prove they exist. He cites the correlation between IQ and brain volume as one example.

Many psychologists would argue that Anderson has it backward. Because there is no consensus in the psychological community as to exactly what intelligence is, it remains a mystical process.[10] IQ, on the other hand, is precisely a psychometric index. It measures an individual's efficiency at one very specific task: answering questions on IQ tests. Not that IQ has nothing to do with smartness or cleverness, whatever those terms may mean. But it cannot say much about intelligence because, again, intelligence has no working definition. Thus any study that claims to have found a connection between anatomy and intelligence has instead found a weak statistical link between an anatomical feature—brain weight is a typical example—and a score on a test of cognitive skills. (It is always worth noting that correlation is not causation, so that even where correlations do exist, they can often be accounted for by a common underlying factor. The frequently cited weak correlation between brain size and IQ could very well be accounted for by a third factor, such as a healthy diet.)

Marian Diamond and Sandra Witelson avoided this mistake by sidestepping the issue of intelligence or genius. IQ scores do not enter into their conclusions. Instead they rely on a localized model of brain function that assigns to Brodmann area 39 and the inferior parietal lobe certain cognitive skills that Einstein seemed to possess in abundance. But there is still a problem.

Witelson describes the inferior parietal lobe as a "secondary association area that provides for cross-modal associations among visual, somesthetic, and auditory stimuli." This is typical of the language of cerebral localization, which assigns a mental faculty such as visual association to a region of the brain by saying that the region "provides for it," or that the faculty is "represented," "encoded," or "instantiated" there. None of these terms have precise definitions. They invoke, rather than describe, physical processes that no one yet understands. As for the psychological processes in question, no one can prove that they have any basis in reality. To say that "visuospatial cognition, mathematical thought, and imagery of movement" are "strongly dependent" on the inferior parietal lobe, as Witelson does, suggests that these are well-established and distinct mental faculties, as opposed to, say, metaphors. But are they?

Pierre Gratiolet asked whether such constructs as sensation, imagination, and memory should be considered distinct faculties of mind, or as different ways of talking about the overriding faculty of "knowing." Paul Broca admitted that the intellectual functions frustrate all attempts at identification. "There is no generic name that encompasses them all," he said, after offering up candidates such as "faculties," "qualities," "sentiments," "penchants," and "passions."[11] Franz Josef Gall, of course, proposed an extensive list of faculties, but even Broca conceded that Gall's program had been a failure.

This turns out to be the subtle flaw in the attempt to locate faculties of mind in regions of the cerebral cortex. What *exactly* (and *exactly* is the key word here) is being localized? If psychology cannot come up with a convincing model of the mind using unambiguous terms to describe its functions, then how is anyone supposed to describe the activity that shows up on the brain scan?

No one today contests the localization of brain function. Pierre Flourens may still have his followers among those who argue for the unity of thought and the interconnectedness of the brain, but even Flourens believed in some distribution of labor among the lobes. Yet what philosophers, psychologists, and cognitive neuroscientists do contest today is the assignment of everyone's favorite mental construct—love for mother, reverence, face recognition, musicality—to one region of the brain.

As Rutgers University philosophy professor Jerry Fodor has noted, "If neuroscience cannot start until psychology gets finished, neuroscience is likely to be in for a long wait." Not surprisingly, the neuroscientists are not waiting, and there, as Fodor points out, lies the practical dilemma:

> On one hand, you can't really ask serious questions about how the brain works if you have no idea what line of work it is in. You need some psychology to prime the neurologist's pump. But, on the other hand, there is not much psychology around that you can rely on if the aspects of mind you wish to study are the "higher cognitive" processes (such as, in particular, thinking).[12]

Another critic of brain imaging, Professor William R. Uttal of Arizona State University, sees the search for higher cognitive processes as a kind of shell game. When it comes to brain functioning, says Uttal, "You can find anything you name."[13] What he means is that it requires little imagination to locate a given faculty of mind somewhere in the brain. He calls this (and his book) *The New Phrenology*. Here is how it works. First name a faculty of mind, any faculty. Godliness, for example, is currently very popular. (On Spurzheim's chart of bumps it was called the faculty of veneration.) Now devise an experiment in which you activate this faculty in the minds of your test subjects. Have them pray or meditate. Then scan their brains using your technology of choice—a PET scan, for example. There will always be a peak of activity. There has to be. (This is dictated by a mathematical law called the extreme value theorem, from calculus.) Adjust the dials, choose the colors, and one part of the brain will light up brighter than the rest. Veneration is thus localized. Keep the sample size small enough, say ten to twelve subjects, and the odds are encouragingly high that, with a little tweaking of the dials, there will be sufficient overlap in the localized zones to allow you to claim, with 95 percent confidence, that there is a "God spot" in the brain, a phrenological module of veneration.

This kind of finding, accompanied as it always is by vivid photographs of richly convoluted and brightly colored hemispheres, always plays well in glossy news magazines. But Fodor and Uttal, among an increasing number of dissenters, have begun to question the merits (and costs) of such studies. On close inspection, they seem to be an instance of technology driving content, while diverting everyone's attention from the man standing behind the curtain.

And yet a tantalizing aspect of brain studies is that just when it looks as if the whole business should be written off as bunk, along comes another positive result that strongly suggests that the brain does indeed contain clues about the mind within, and that the phrenologists may have been right.

AT THE DAWN of the new millennium, a research group from the University of Heidelberg conducted a study in which they compared the brains of musicians to nonmusicians by measuring anatomical differences in their temporal lobes through computer imaging, and by timing their brains' responses to musical stimuli. The results, published in the prestigious journal *Nature Neuroscience* in the summer of 2002, were conclusive, and not unexpected. They found that a gyrus in the primary auditory cortex (Heschl's gyrus) is 2.3 times as big and twice as active in the brains of professional musicians as in nonmusicians. Because Heschl's gyrus is not supposed to grow after the age of nine, the verdict is clear—start playing an instrument before then if you want to be in the orchestra.

Studies like this seem to indicate that there must be something to phrenology, not in the bumps in the skull but in the convolutions of the cerebral hemispheres. In localized zones devoted to very specific skills, more does seem to be better. At the very least, greater development coincides with greater performance capacity. But the strained wording of the previous sentence hints at the elusiveness of the concept. How should any of this be interpreted?

In its summary of the Heidelberg study, the *New Scientist* magazine showed how not to do it. According to their headline: "If you're not musical now, you never will be."[14] The study itself said nothing of the sort. Scientists tend to be far more cautious than journalists. The Heidelberg group was careful not to let a term like "musical" dangle in the breeze. They divided their control group into three clearly defined subgroups—nonmusicians, amateur musicians, and professional musicians. They made no suggestion as to the futility of taking up a musical instrument in adulthood. They would not conjecture about the difference that practice makes. They concluded only that the structure and function of Heschl's gyrus has a strong correlation with musical aptitude. "The question remains, however, whether early exposure to music or a genetic predisposition leads to the functional and anatomical differences between musicians and non-musicians." In other words, practice can make perfect, and it may do so by building a bigger Heschl's gyrus. Or it may not.

The finding was not a new one. Brain morphologists of a century earlier had noted (but not measured) overdevelopment of the Heschl gyrus in brains of musical prodigies and professional musicians. In 1906, the German anatomist Siegmund Auerbach found a redundancy (to use Spitzka's term) in the auditory association areas in the brains of the conductor-composers Hans von Bülow and Felix Mottl, the music teacher Naret Koning, the pianist Goswin Sökeland, the violinist Rudolpf Lenz, and Bernhard Cossmann (who was known as the "Joachim of the violoncello").[15] At last, a well-designed study had confirmed and given meaning to case studies from Spitzka's era. Perhaps the sacrifice of the brain donors had been redeemed after all.

But again, not so fast. Before the Heidelberg findings can be accepted, they have to be replicated, which will take time. Ideally, a longitudinal study could track the brain development of subjects from birth into adulthood using a sample large enough to assure a sufficient number of musicians and nonmusicians in the final tally. The logistics of such an undertaking are daunting, the benefits marginal. It is unlikely to be done soon. The same obstacles discourage attempts to isolate other talents, to pinpoint brain-anatomical differences between mathematicians and a control group of innumerates, or between professional chess players and nonplayers, for example. Given the lack of urgency, the absence of either a life-and-death imperative or a profit motive for pharmaceutical companies, progress in this area can be expected to be glacial, and, consequently, the least hint of progress will only fan the flames of hype. Until a firm connection between human performance and inherited anatomical features of brains is demonstrated beyond a reasonable doubt (a prospect that seems increasingly remote), conjectures based on shaky evidence will breed misconceptions at best, and harebrained eugenic interventions at worst.

It is still not clear which comes first—whether musical exposure leads to an enhanced Heschl gyrus, or whether a natural endowment in the temporal lobes facilitates a musical career. But the implications are troubling. What if one's musical potential lies entirely in the genes and can be revealed by a brain scan at birth? Should children then be screened and assigned according to brain structure to special groups—the music group, the math group, the art group? Should those with deficient Heschl gyri be discouraged from taking up a musical instrument? Should the determination of what constitutes musicality be relegated to computers?

In his essay on Broca's brain, Carl Sagan invoked the Faust legend (which is, in essence, the Frankenstein legend) as the embodiment of the lurking fear that some things are not meant to be known, that nature holds secrets that men unearth at their peril. He concludes that such fears are worth fighting, as long as inquiries are conducted openly, as long as results are shared. If the public is scientifically literate, according to Sagan, the potential for abuses is minimized. But how realistic is that when the subject of study is as recondite as the human brain, and the specimen in question is an Einstein?

Chapter 15

The New Phrenologists

THE RESPONSE TO Sandra Witelson's report on Einstein's brain was predictable, and surpassed the furor that attended Marian Diamond's glial cell study. But it too was limited to the popular press. In the professional neuroscientific community, it failed to make a dent. Which is to say that it received almost no peer review. This was only fitting, because as scientific as it may have sounded, "The Extraordinary Brain of Albert Einstein" was not really science at all.

Writing for the *New York Times*, MIT cognitive psychologist Steven Pinker praised Witelson's paper as an "eloquent study" that was "consistent with the themes of modern cognitive neuroscience."[1] Pinker, the author of such works as *The Language Instinct, How the Mind Works,* and *The Blank Slate,* is an evolutionary psychologist, which is to say he supports the theory of innate mental faculties and supporting cortical structures. In an op-ed piece entitled "His Brain Measured Up," he warmed to Witelson's broad description of the association areas of Einstein's brain, and went so far as to say that Witelson's study "confirms that the brain is a modular system comprising multiple intelligences, mostly nonverbal."

That's one way of looking at it, although Witelson's study could just as easily be dismissed, to borrow Auguste Comte's words, as "vulgar charlatanism," or as mere idolatry. Certainly, if the short history of anatomical studies of famous brains has demonstrated anything, it is the impossibility of contemplating the brain of a genius as a mere object. The label on the jar always gives the game away. Time and again, the experts have allowed themselves to be swayed by their preconceptions, and reacted to the label rather than the jar's contents.

Pinker was happy to report that Sandra Witelson did not argue against the evidence, but in his enthusiasm he seemed to forget that the object of his praise was a description of a few photographs of a brain, rather than a real brain, and that her study did not confirm the modularity of brain function, or much of anything else. In a sudden shift from overstatement to understatement, Pinker conceded that "we are not going to work out the wiring diagram in Einstein or anyone else any time soon," and so deflated any expectations that might have been raised by the title that had been tacked onto his piece. Einstein's brain may have measured up, but to what?

ACCORDING TO HIS many admirers, Albert Einstein was a god. Legend has it that he was a poor student who flunked math, and then invented relativity while daydreaming in a Swiss patent office. Einstein himself did nothing to cultivate this myth, but he was aware of it and seemed bothered by it. He made a point of saying that for every hundred ideas that came to him, ninety-nine were wrong.

In his youth, Einstein was an intensely independent student who rebelled against the standard academic curriculum, and followed his own interests. He surrounded himself with friends of like mind, and together they discussed philosophical and scientific ideas. Nothing came to him without effort. A concentrated period of deep study and hard work led to his first breakthrough. A lifetime of deep concentration probably affected the growth and structure of his brain, and in theory, its surface patterns contained a clue to the life of his mind. But the studies of his brain published thus far suggest a simpler explanation, if not a simplistic one: not only was Einstein unlike anyone else, he was not even human.

In his *Times* article, Pinker made this explicit when he wrote that "the difference between the inferior parietal lobes of Einstein and of *us mortals* is not subtle" (emphasis added). Pinker can be excused for indulging in a bit of poetic license, and he was careful to add that Einstein's mental gifts "surely lie in the microcircuitry formed by millions of synapses in many parts of his brain." But the point was made. Faced with the legacy of an Einstein, many competent scientists are apt to lose their objectivity, and temporarily exchange their scientific creed for a more mystical one.

It is a great irony that the triumph of scientific materialism, the very movement that freed science from the yoke of religion, should have erected in its place a religion of its own, a religion in which intellectual giants like Einstein and Gauss took the place of saints, in which autopsies took the place of last rites, in which brains became sacred relics, and brain studies served up hagiography. In this new religion, genius remains an elusive ideal, the intellectual equivalent of perpetual motion. The very idea that geniuses can bestride the heavens without working up a sweat seems to violate the law of conservation of energy, but this is precisely how materialist scientists continue to speak, even though it violates their core belief. Edward Anthony Spitzka succumbed to this peculiar confusion of objectivity and veneration when he wrote: "It is not enough merely to admire the genius of an Archimedes, a Newton, a Michel Angelo or a Bacon; we wish to know how such men of 'brains' were capable of their great efforts of the intellect and what gave them the capacity for doing great things, as it were, 'without taking pains.' "[2]

Given the choice between the myth of the natural genius and the sobering reality of hard work, popular culture usually opts for the myth. It is a far more palatable scenario; in fact, it is *the* Frankenstein scenario. The idea that nature, or perhaps God (and every now and then a mad scientist) endows special people with brains that allow them to do great things with no effort makes few demands on an audience, and allows one of the staples of ancient myth—the mind as spirit—to merge with the modern concept of the mind as brain.

A century and a half after Rudolf Wagner preserved the brain of the great Carl Gauss, Spitzka's wish remains unfulfilled. Brain scientists have not only failed to localize the higher thought processes, but they still cannot distinguish the brain of a Gauss or an Einstein from that of a depraved criminal. Even Pinker concedes that "no one can claim to have explained Einstein's genius. For all we know a person with big inferior parietal lobules could just as easily become a great home builder or billiard's shark." Or a mass murderer.

It is hardly the case that all of contemporary brain science is a fraud. In every other area, in brain pathology, neurochemistry, neurosurgery, the treatment of tumors, on all fronts excepting this one, scientists have made astounding progress over the last century. By comparison, the search for the anatomical substrate of intelligence has succeeded in creating and nurturing a discouraging idea. In place of any real insight on the subject, it has propped up the myth of the born genius.

IN THE ORIGINAL sense of the word, a genius (or genie) was a minor god, and everybody had one. In ancient Greece, where household and personal gods manipulated the details of everyday life, the genie served a useful purpose. He (or she) arrived on the scene at the moment of birth, guided an infant through the four stages of life, shaped its character along the way, and then departed at death in order to start the cycle anew. Some genies happened to be better connected than others, and a very few, in all of recorded history, had direct access to the sublime. This explained geniuses of the highest order, like Wolfgang Amadeus Mozart, whose middle name suggests what many people still believe—that he was favored by God.

As scientific materialism took hold in the 1800s, the idea of genius descended from its Olympian heights. Prior to the time of, say, Byron, most references to genius were in the possessive; one could speak of Shakespeare's genius, Dante's genius, or the genius of Michelangelo, and preserve the original sense of the word. It was less common, if not unheard of, for someone to say, "This man is a genius!" But by the late 1800s, the attribution of genius had become common (today it has become an almost meaningless cliché), and it carried a very specific connotation. To call someone a genius meant that his or her intellect was part of that person's essential makeup rather than an attendant spirit; genius then emerged as a quality of the mind that could be attributed to the organization of a superior brain. In the process, materialist scientists transformed genius into brain power, and once they did, it was only natural for them to try to measure it.

From the start, the search for the anatomical basis of genius, of criminality, of insanity, and of specific talents was plagued by errors of oversimplification, of reading too much significance into unusual cases, errors due to "the personal equation of the investigator," to the inadequacies of language, and finally, to subservience to schools of thought, including phrenology, positivism, dialectical materialism, Cartesian dualism, and biological determinism. Researchers like Gall, Broca, Lombroso, Spitzka, and Vogt routinely began their investigations by assuming the truth of what they wanted to show, then tried to find the data to back it up. This was, as Stephen Jay Gould put it, a case of "hope dictating conclusion,"[3] and as the downwardly spiraling reputations of these men showed, it was a high-stakes and risky game to play.

When asked whether she had any regrets about her study of Einstein's brain, Marian Diamond said no, adding, "This is what scientists do!"[4] Which is true. Scientists are in the business of chasing down possibilities. But she was overlooking an important point. Scientists usually chase down possibilities in order to rule them out. Which is to say that they are trained to be critical of their own ideas, to make every effort to disprove them before announcing them to the world. This is what the English philosopher Francis Bacon had in mind when he wrote: "If a man will begin with certainties, he will end in doubts, but if he will be content to begin with doubts, he shall end in certainties."[5] It was what Charles Darwin had in mind when he set aside his *On the Origin of Species* in 1846 in order to devote eight grueling years to the classification of barnacles.[6] Darwin did not do this out of curiosity, but to prove to himself that his evolutionary theory was right—in other words, to rule out every possibility that he was wrong.

THE NOTION THAT there is such a thing as a mathematical brain, a musical brain, a genius brain, or a criminal brain has served a useful purpose. It has driven the plot of many popular films and novels, and provided some entertaining urban folklore. Maybe there really is a "God spot" or a "math spot," maybe there are left-brained and right-brained people, maybe there is a criminal type. If these things have not been proven, at least they have not been ruled out.

But the downside of unsubstantiated phrenological claims often goes unnoticed. The notion that the limits of a mind's potential are fixed by the shape of the brain that houses it offers little room for optimism. Ironically, the most redeeming aspect of the "practical" phrenology of the Fowlers was its insistence that anyone can improve their mind by exercising their brain on a regular basis. Human brains are indeed highly adaptable. Modern clinical neurology has proven that this is very much the case.

In *The Mismeasure of Man*, Stephen Jay Gould wrote, "I do confess to a warm spot in my heart for the phrenologists . . . for they were philosophically on the right track," by which he meant that they "celebrated the theory of richly multiple and independent intelligences."[7] But the Fowlers seemed to be on the right track for another, purely pragmatic, reason. As misguided as their elaborate charts may have been, they held out a hope that was not entirely unfounded. They encouraged people to do something about their lives and suggested constructive ways to go about it.

The new phrenologists are more fatalistic. They have appropriated the phrenological conceit that mental deficiencies can be diagnosed, while discarding the quaint notion that something can be done about them. By focusing exclusively on the most distinctive features of special brains—Einstein's wayward Sylvian fissures, Gauss's tortuous convolutions, Lenin's swollen pyramid cells—they have succeeded in reducing a multifaceted phenomenon, the mind, to a single dimension—an odd feature of the brain.

According to Messrs. Fowler and Wells, you can do something about the brain you were given. This was music to Walt Whitman's ears. But the new phrenologists are saying, in essence: No, you can't. To the student who insists, "I'm just no good at math," they reply, "You're probably right." To the hopeful young violinist whose Heschl gyrus is too small they say, "Give up." To the child whose Sylvian fissure does not take the same shortcut as Einstein's, they say, "You're no Einstein."

What the new phrenologists have failed to realize is that the exaltation of some brains always comes at the expense of others. According to Sandra Witelson, Einstein was "one of the intellectual giants of recorded history," which would make his brain one of the best thinking instruments of all time. The same could be said of Gauss with even more justification, but no one bothers anymore. The irony is that the claim for Einstein's brain is made with the imprimatur of science, yet its premise is undermined by the same logical errors that plagued Spurzheim's phrenology.

WHEN ÉDOUARD SÉGUIN opened his school for mentally retarded children in 1834, he single-handedly challenged a well-entrenched school of thought that claimed that so-called "idiot" children were unteachable. This was the working hypothesis of his peers, and, during seven years of labor, Séguin amassed enough evidence to convince the French Academy of Science to reject it. His teaching philosophy may have been overly optimistic, but optimism was essential to the job. He may not have understood brain anatomy, brain physiology, or the limits of his undertaking, but no one could argue with his results. Today, Séguin's spirit lives on in the Montessori method, and it is carried on by patient teachers who help mentally retarded and autistic children to lead better lives. Such is the power of an idea, and such is the adaptability of the brain.

Séguin did not bequeath his own brain to his friend Edward Charles Spitzka in the hope that it would confirm his genius, or even his eminence. He believed in the potential of science to improve lives, and he assumed that his brain might contribute to the understanding of the organ of the mind. He was an atheist and a freethinker himself, but he did not know the direction that freethinking would take, or what would become of his brain. He did not know that brain anatomical studies would be used to convince people of their limitations, and to argue that certain classes of people should be removed from the breeding population. Nor did he suspect that collections of brains would lead to the classification of degenerate types, and that all of this would come to pass in the name of science, with little science behind it.

THE BRAIN, SAID Franz Josef Gall, is the exclusive organ of the mind. It is composed of modules that give rise to its varied powers and faculties. The basis for these dispositions is innate, and it depends on the brain's gross and cellular structure. This remains the reigning paradigm of modern neuroscience and the basis for studies of famous brains. But it is not a complete paradigm in the sense that Thomas Kuhn intended the term. Unlike the Copernican view of the sun as the center of the solar system, or Einstein's relativistic view of time, space, and matter, it has not been verified by observation and measurement. It seems unlikely that it ever will be.

As early as 1840, Auguste Comte saw that something was missing, that in his three-tiered scheme for the advancement of knowledge, the science of the mind was stuck somewhere between the religious and the metaphysical stages. He believed that phrenology could take psychology to the next level, to the status of a pure science. But it hasn't. It has failed to produce working definitions of the mental faculties, particularly of intelligence, that do not rely on metaphor or on mere consensus.

If there was something truly distinctive in the brains of great men, great women, depraved hoodlums, or murderers, it would have been discovered by now. But nothing has turned up. Not that Henry Frankenstein had it right when he suggested that a preserved brain really is just a piece of dead tissue. There have been some suggestive findings, particularly in the brains of mathematicians and musicians, but not enough to validate Dr. Waldman's claim at the beginning of the classic film: No one can look at a brain and tell what sort of person inhabited it. Nor has anyone discovered a scientific basis for judging the superiority of one mind over another, or of one brain over another (short of true pathology).

What Oskar Vogt's critics, the Harvard neuroanatomists Percival Bailey and Gerhard von Bonin, noted in 1951 remains true today. The safest conclusion still appears to be "that all human brains look essentially alike." There are differences, of course, but such differences do not yet allow one brain shape or cytoarchitectonic map to be preferred over others. The only reasonable assumption is that at the outset of life, any healthy brain has a potential that is essentially infinite. For the time being, this is the only legitimate working hypothesis available, and it is easily the most constructive one. Despite occasional claims to the contrary, no one has managed to dislodge it.

Acknowledgments

The author wishes to thank the archivists, curators, and researchers who opened their facilities, shared their expertise, or provided helpful suggestions. These include:

Elizabeth Allen, Hunterian Curator, Royal College of Surgeons of England
Maria Asp, Royal Swedish Academy of Sciences, Stockholm
Beth Bensman, Thomas Jefferson University Archives
Barbara Finlay, Department of Psychology, Cornell University
Ed Folsom, University of Iowa
Andreas Frewer, Institute for Medical Ethics and the History of Medicine, Göttingen
Gunnar Grant, Department of Neuroscience, Karolinska Institute, Stockholm
Duane Haines, University of Mississippi Medical Center
Nina Long, Curator, The Wistar Museum Collection
Philippe Mennecier, Musée de l'Homme, Paris
Roger Saban, Muséum National d'Histoire Naturelle, Paris
Olaf Stroh, Medicinhistoriska museet, Stockholm
Paolo Tappero, Museo di Antropologia Criminale, Torino

William R. Uttal, Arizona State University
Axel Wittmann, Director, Göttingen Observatory
Gretchen Worden, The Mütter Museum, Philadelphia
College of Physicians
Karl Zilles, C. and O. Vogt Institute for Brain Research,
Heinrich Heine University, Düsseldorf

With special thanks, for their gracious hospitality, to Michael Hagner, Max Planck Institute for the History of Science, Berlin; to Jochen Richter, independent scholar of Berlin; and most of all to Ilya Victorov of the Institute of the Brain in Moscow.

Literal translations and interpretation of German texts were insightfully provided by Volker Ecke. Translations from the Russian were adroitly handled by Nina Korotkova. Responsibility for the final rendering of all translations lies entirely with the author.

Appendix

Paris Collection

Musée de l'Homme—Société Mutuelle d'Autopsie

Jules Assezat (1832–1876), writer
Louis Asseline (1829–1878), journalist
Leon Gambetta (1838–1882), statesman
Adolphe Bertillon (1821–1883), anthropologist and criminologist
Auguste Coudereau (1832–1882), physician
Eugène Véron (1825–1889), writer
Abel Hovelacque (1843–1895), writer
——— Chèvre (unknown)
José-Maria Guardia (1830–1897)
Gabrielle de Mortillet (1821–1898), anthropologist
Joseph Bouny (unknown), jurist
Charles Letourneau (1831–1902), anthropologist
Jean Vincent LaBorde (1830–1903), physiologist and anthropologist

——— Lavolloy (unknown)
(Madame) Leblais (unknown), educator

MUSÉE DUPUYTREN

Paul Broca (1824–1880), surgeon, anatomist, anthropologist
Lelong, aphasic patient
Leborgne, aphasic patient

Moscow Collection

INSTITUT MOZGA

Vladimir Ilyich Lenin (1870–1924), revolutionary
Joseph Stalin (1879–1953), Soviet leader
Sergei Eisenstein (1898–1948), theater and film director
Ivan Pavlov (1849–1936), physiologist
Vladimir Mayakovsky (1893–1930), poet
Andrei Sakharov (1921–1989), physicist
Maxim Gorky (1868–1936), writer and statesman
Clara Zetkin (1857–1933), German labor figure, member of the Comintern
M. M. Ippolitova-Ivanovna (1859–1935), composer, conductor, teacher
A. V. Lunacharsky (1875–1933), statesman, writer, critic, art historian
V. R. Menzhinsky (1874–1934), Bolshevik Party official
Alexander Bogdonov (1873–1928), economist, philosopher, biologist
Ivan Michurin (1855–1935), biologist, father of Soviet genetics

Konstantin Tsiolkovsky (1857–1935), scientist, founder of Soviet astronautics
Ivan I. Skvortsov-Stepanov (1870–1928), historian, economist
Sen Katayama (1859–1933), Japanese Comintern official
Henri Barbusse (1873–1935), French writer
Andrei Bely (1880–1934), poet and novelist
Eduard Bagritsky (1895–1934), poet
Leonid Sobinov (1872–1934), opera singer
Sergei Kirov (1866–1934), Soviet political leader
Mikhail Kalinin (1875–1946), politician

Tokyo Collection

University of Tokyo Medical Museum

Sekizen Arai, Zen buddhist priest
Osachi Hamaguchi, twenty-seventh prime minister of Japan
Homei Iwano, novelist
Naozo Ichinohe, astronomer
Taro Katsura, eleventh, thirteenth, and fifteenth prime minister of Japan
Teiichi Kashimura, medical scientist
Noboru Kanai, imperial household medical doctor
Yoshinao Kozai, president of Imperial University
Shuzou Kure, professor of surgery
Yasuko Miyake, female writer
Chomin Nakae, society critic
Soseki Natsume, novelist and essayist
Naito Meisetsu, haiku poet
Tadayori Nakagawa, art historian
Seiichi Terano, professor of shipbuilding
Kanzou Uchimura, Christian evangelist

Kaigyoku Watanabe, priest of the Jodo sect
Katsusaburo Yamagiwa, professor of pathology
Kenjiro Yamakawa, physicist, president of Imperial
University
Yoshitomi Mishikida, professor of ethics
and one hundred others

Philadelphia Collection

Wistar Institute of Anatomy and Biology

Dr. Milton J. Greenman (1866–1937), anatomist, third
director of the institute
Dr. Henry H. Donaldson (1857–1938), physiologist
Dr. Harrison Allen (1841–1897), comparative anatomist
Prof. William Keith Brooks (1848–1908), zoologist
Prof. Edward Drinker Cope (1840–1897), ichthyologist,
paleontologist
Prof. Joseph Leidy (1823–1891), natural scientist
Dr. Philip Leidy (1839–1891), physician
Dr. William Pepper (1843–1898), university provost
Prof. James Tyson (1841–1919), professor of pathology and
morbid anatomy
Dr. James William White (1850–1916), professor of surgery
Dr. E. H. Dunn, physiologist
Dr. George Arthur Piersol (1856–1924), professor of anatomy
Prof. Edward Sylvester Morse (1838–1925), professor of
zoology, expert on Japanese ceramics
Prof. G. Stanley Hall (1846–1924), experimental psychologist,
president of Clark University
Dr. William W. Keen (1837–1932), surgeon

Gen. Isaac Jones Wistar (1827–1905)
Sir William Osler (1849–1919), pathologist
Prof. J. S. Haldane (1860–1936), physiologist
Prof. H. Wade Hibbard (1863–1929), professor of mechanical engineering
Dr. Alfred Stengel (1868–1939), physician
H. Pillsbury

Spitzka's List of the Top 100 Brain Weights, circa 1912

1. Ivan Turgenev, poet and novelist	2,012 grams
2. Georges Cuvier, naturalist	1,830
3. E. H. Knight, physicist	1,814
4. [unnamed theologian from Freiburg]	1,800
5. John Abercrombie, physician	1,786
6. Benjamin Butler, lawyer and general	1,758
7. Edward Olney, mathematician	1,701
8. Herman Levi, composer	1,690
9. William Makepeace Thackeray, novelist	1,658
10. Rudolph Lenz, composer	1,636
11. John Goodsir, anatomist	1,629
12. Hosea Curtice, mathematician	1,612
13. C. G. Atherton, U.S. senator	1,602
14. W. V. Siemens, physician	1,600
15. George Brown, editor	1,596
16. A. Konstantinov, literateur	1,595
17. R. A. Harrison, chief justice of Canada	1,590
18. F. B. W. V. Herman, economist	1,590
19. J. K. Riebeck, philologist	1,580

20. Hans Buchner, hygienist	1,560 grams
21. Caspar Spurzheim, phrenologist	1,559
22. Lavollay, publicist	1,550
23. Edward D. Cope, paleontologist	1,545
24. G. McKnight, physician and poet	1,545
25. Harrison Allen, anatomist	1,531
26. J. Y. Simpson, physician	1,531
27. Pierre Dirichlet, mathematician	1,520
28. C. A. DeMorny, statesman	1,520
29. Daniel Webster, statesman	1,518
30. Lord John Campbell, Lord Chancellor of England	1,517
31. Chauncey Wright, philosopher	1,516
32. M. Schleich, writer	1,503
33. Thomas Chalmers, theologian	1,503
34. Garrick Mallery, ethnologist and soldier	1,503
35. Edward C. Séguin, neurologist	1,502
36. Napoléon III	1,500
37. K. H. Fuchs, pathologist	1,500
38. Louis Agassiz, naturalist	1,495
39. Carlo Giacomini, anatomist	1,495
40. De Morgan, mathematician	1,494
41. Carl F. Gauss, mathematician	1,492
42. Letourneau, anthropologist	1,490
43. John Wesley Powell, geologist	1,488
44. K. V. Pfeufer, physician	1,488
45. Wulfert, jurist	1,485
46. Paul Broca, anthropologist	1,484
47. Gabriel de Mortillet, anthropologist	1,480
48. P. Aylett, physician	1,474
49. Lord Francis Jeffrey, jurist	1,471
50. Louis Asseline, journalist	1,468
51. M. D. Skobelev, general	1,457

52. C. H. E. Bischoff, physician	1,452 grams
53. Hugo Gylden, astronomer	1,452
54. LaMarque, general	1,449
55. F. R.V. Kobell, geologist and poet	1,445
56. Mihalovicz, embryologist	1,440
57. Hermann von Helmholtz, physiologist	1,440
58. Guillaume Dupuytren, surgeon	1,437
59. Per Adam Siljestrom, physicist	1,432
60. Franz Schubert, composer	1,420
61. A. T. Rice, diplomat	1,418
62. J. E. Oliver, mathematician	1,416
63. Melchior Meyr, philosopher	1,415
64. Joseph Leidy, morphologist	1,415
65. Philip Leidy, physician	1,415
66. George Grote, historian	1,410
67. Nussbaum, surgeon	1,410
68. W. J. McGee, anthropologist	1,410
69. John Huber, philosopher	1,405
70. Charles Babbage, mathematician	1,403
71. Jules Assezat, journalist	1,403
72. C. V. Kupffer, anatomist	1,400
73. Louis-Adolphe Bertillon, anthropologist	1,398
74. Fr. Goltz, physiologist	1,395
75. Auguste Coudereau, physician	1,390
76. William Whewell, philosopher	1,389
77. Henry Wilson, vice president	1,389
78. N. Rudinger, anatomist	1,380
79. S. Zilagyi, statesman	1,380
80. H. T. V. Schmid, writer	1,374
81. A. A. Hovelacque, anthropologist	1,373
82. Theodor Bischoff, anatomist	1,370
83. K. F. Hermann, philologist	1,358
84. Justus Leibig, chemist	1,352

85. V. Schagintweit, naturalist	1,352 grams
86. J. P. Fellmerayer, historian	1,349
87. John Hughes Bennett, physician	1,332
88. Max Von Pettenkofer, chemist	1,320
89. Seizel, sculptor	1,312
90. J. G. Kolar, dramatist	1,300
91. R. E. Grant, astronomer	1,290
92. Walt Whitman, poet	1,282
93. Édouard Séguin, psychiatrist	1,257
94. Friedrich Tiedemann, anatomist	1,254
95. V. La Saulx, philologist	1,250
96. LaBorde, physiologist	1,234
97. L. V. Buhl, anatomist	1,229
98. J. F. Haussmann, minerologist	1,226
99. B. G. Ferris, jurist	1,225
100. Franz Josef Gall, anatomist	1,198*

*a transcription error; it should read 1,312 grams

From Edward Anthony Spitzka, "A Study of the Brain of the Late Major J. W. Powell," *American Anthropologist* 5 (1903), 595–96.

Notes

Introduction

1. Shelley, introduction to *Frankenstein* (London: H. Colburn and R. Bentley, 1831), p. viii.
2. In Darwin's *The Temple of Nature,* he refers to a "vorticella or wheel animal," and not to pasta. It was a simple misunderstanding on Shelley's part.
3. Shelley, *Frankenstein,* p. x.
4. Sarah Lyall, "Inquiry Shows British Scientists Took Brains Without Families' Consent," *New York Times,* 13 May 2003, p. A8.
5. For more on McLean Hospital, see Raja Mishra, "What Makes Us Tick?" *Boston Globe,* 9 January 2001, pp. F1–3.
6. Brigitte Greenberg, "Med Student Unlocks Surgical Pioneer's Trove," *Philadelphia Inquirer,* 3 March 1996.
7. Sagan, *Broca's Brain,* p. 7.

Chapter 1: The Most Complex Object in the Universe

1. Clark and O'Malley, *Human Brain and Spinal Cord,* p. 385.
2. Finger, *Minds Behind the Brain,* p. 35.
3. Clark and O'Malley, *Human Brain and Spinal Cord,* p. 403.

Chapter 2: Descartes

1. Quoted in Finger, *Minds Behind the Brain*, p. 136.
2. Perier, *Bulletin de la Société d'Anthropologie de Paris*, 4 April 1861, p. 224.
3. Finger, *Minds Behind the Brain*, p. 29.
4. Descartes, *Discourse on Method*, parts II and IV.
5. Descartes, *Treatise on Man*, p. 141.
6. Gilbert Ryle, *The Concept of Mind*, chapter 1.
7. Winter, *Podromus of Nicolaus Steno's Dissertation*, pp. 227–31.
8. Cutler, *Seashell on the Mountaintop*, p. 39.
9. Steno, "Discours sur l'anatomie du cerveau," p. 200.
10. Cutler, *Seashell on the Mountaintop*, p. 42.
11. Ibid., p. 91.
12. Martensen, "When the Brain," p. 23.
13. Kuhn, *Structure of Scientific Revolutions*, p. 10.
14. Ibid., p. 17.
15. Jeannerod, *Foundations*, p. 10.

Chapter 3: Gall

1. For Mark Twain on phrenology, see his *Autobiography*, chapter 13.
2. Ackerknecht and Vallois, *Franz Josef Gall*, p. 8.
3. Ibid.
4. Baron d'Holbach, *The System of Nature*, volume 1, chapter 7.
5. Hartley, *Physiognomy and the Meaning of Expression in Nineteenth-Century Culture*, p. 34.
6. Finger, *Minds Behind the Brain*, p. 133.
7. Gall's four points can be found in his last work, *Sur les fonctions du cerveau* (Paris, 1822), p. vi.
8. Bailey, "Seat of the Soul," p. 423.
9. Finger, *Minds Behind the Brain*, p. 134.
10. Macmillan, *An Odd Kind of Fame*, p. 146.
11. Ambrose Bierce, *Devil's Dictionary*, p. 252.
12. Williams, *Gray's Anatomy*, p. 869.
13. *Encyclopedia of Philosophy*, vol. 2, p. 173.
14. Mill, *The Correspondence of John Stuart Mill and Auguste Comte*, p. 58.
15. François Leuret, *Du Traitement Moral*, pp. 49–50.
16. Berlin, "Historical Inevitability," pp. 41–43.
17. Mill, *The Correspondence of John Stuart Mill and August Comte*, p. 80.
18. Comte, *Cours de philosophie positive*, chapter 6, book 4.

Chapter 4: Byron

1. Marchand, *Byron*, pp. 1224–25.
2. Finger, *Minds Behind the Brain*, p. 34.
3. Iserson, *Death to Dust*, p. 83.
4. For more on the history of dissection, see Iserson, *Death to Dust*; and James Moore Ball, *The Body Snatchers* (New York: Dorset Press, 1989).
5. Spitzka, "Study of the Brains of Six Eminent Scientists," p. 177.
6. Bailey, "Seat of the Soul," p. 421.
7. Jacobsen, *Foundations of Neuroscience*, p. 183.
8. Jeannerod, *Brain Machine*, p. 6.
9. For a brief history of preservative methods, see Williams, *Gray's Anatomy*, 37th edition, p. 869.
10. In his article "Brain: Methods of Removing and Preserving," for the *Reference Handbook of the Medical Sciences*, Burt Green Wilder gives a brief overview of how *not* to preserve a brain.
11. Bailey and von Bonin, *Isocortex of Man*, p. 16.
12. Todd, "Liter and a Half of Brains," p. 124.
13. *Oeuvres de Vicq-d'Azyr* (Paris, 1805), vol. 4, p. 208.

Chapter 5: Gauss

1. Piazzi's data was published in the September 1801 issue of von Zach's monthly correspondence *(Monatliche Correspondenz)*, p. 280.
2. C. Vogt, *Köhlerglaube*, p. 17.
3. Wagner, *Conversations with Gauss*, 19 December.
4. C. Vogt, *Köhlerglaube*, p. xxxiv.
5. W. K. Bühler, *Gauss*, p. 154.
6. Gould, *Mismeasure of Man*, pp. 401–12.
7. Ibid., p. 406.
8. Wagner, *Vorstudien* (1860), p. 90.
9. Carl Vogt, *Lectures on Man*, p. 84.

Chapter 6: Broca

1. Schiller, *Paul Broca*, p. 135.
2. The transcript of the Broca-Gratiolet debate is contained in the *Bulletin de la Société d'Anthropologie* for the year 1861.

3. *Bulletin de la Société d'Anthropologie,* 20 December 1861, p. 67.
4. Hecht, *End of the Soul,* p. 57.
5. *Bulletin de la Société d'Anthropologie,* 21 February 1861, p. 78.
6. Young, *Mind, Brain and Adaptation in the Nineteenth Century,* pp. 135–36.
7. The Bouillaud and Dax papers are: Marc Dax, "Lésion de la moitié gauche de l'encéphale coïncidant avec l'oubli des signes de la pensée" *Gazette hebdomodaire* 2nd series, 2 (1865) :259–60; and J. B. Bouillaud, "Recherches cliniques propres à démontrer la perte de la parole correspond à la lesion des lobules antérieurs . . ." *Archive Générale de Médecine* 3, no. 8 (21 February 1825): 25–45.
8. *Bulletin de la Société d'Anthropologie,* 21 February 1861, p. 76.
9. Schiller, *Paul Broca,* p. 268.
10. Ibid., p. 139.
11. The account of Broca's presentation of the brain of Tan is in the *Bulletin de la Société d'Anthropologie,* 4 April 1861.
12. Schiller, *Paul Broca,* p. 192.
13. For Pierre Marie on Tan's brain, see ibid., p. 208.
14. Gratiolet, *Memoire,* pp. 101–102.
15. *Bulletin de la Société d'Anthropologie,* 21 March 1861, p. 196.
16. Ibid., 18 April 1861, p. 270.
17. Ibid., p. 238.
18. Dally's speech, ibid., 16 January 1862, p. 51.
19. Hecht, *End of the Soul,* pp. 13–15.
20. The subject of an autopsy society was first broached at the 19 October 1876 meeting of the Société d'Anthropologie. *Bulletin,* series 2, volume 11.
21. For more on de Mortillet, see Hammond, "Anthropology as a Weapon," pp. 119–22; and Hecht, *End of the Soul,* pp. 136–39.
22. Hecht, *End of the Soul,* p. 13.
23. Quoted by Jean d'Echérac, "André Lefèvre," *Revue mensuelle de l'Ecole d'Anthropologie de Paris* 11 (1904): 383–86.
24. *Bulletin de la Société d'Anthropologie,* 18 April 1878, p. 164.
25. Ibid., 5 April 1872, p. 164.
26. Ibid., 21 July 1887, p. 561.
27. Bastian, *The Brain as Organ,* 370–71.

Chapter 7: Lombroso

1. Joseph Conrad, *The Secret Agent,* p. 47.
2. Wolfgang, *Pioneers in Criminology,* p. 183.
3. Gibson, *Born to Crime,* p. 19.
4. Wolfgang, *Pioneers in Criminology,* p. 182.

5. Marx and Engels, *The Holy Family*, chapter 5, part 1.
6. Pick, *Faces of Degeneration*, p. 118.
7. Gould, *Mismeasure of Man*, p. 153.
8. Gibson, *Born to Crime*, p. 17.
9. Naumov, *Rossiiskoe Ugolovnoe Pravo*, p. 544.
10. Bischoff, *Das Hirngewicht des Menschen, Eine Studie* (Bonn: 1880).
11. Benedikt, *Anatomical Studies*, p. 157.
12. Osler's review of Benedikt appears in "On the Brains of Criminals."
13. Osler, "On the Brains," p. 398.
14. For an in-depth discussion of recapitulation and its long history, see Gould, *Ontogeny and Phylogeny*.
15. For Lombroso reapportioning crime rates to reflect degeneration, see Wolfgang, *Pioneers in Criminology*, p. 188.
16. Lombroso, *Man of Genius*, p. 91.
17. Ibid., p. 9.
18. The story of Lombroso's interruption at Giacomini's postmortem presentation is told in Francesco Loreti, *Carlo Giacomini: Contributo alla Storia dello "Studio" Anatomical della Università di Torino* (Torino: Accademia Delle Scienze, 1963), p. 15.
19. For Lombroso on Darwin and Michelangelo, see *The Man of Genius*, pp. 353–58.
20. Wolfgang, *Pioneers in Criminology*, p. 207.
21. Pick, *Faces of Degeneration*, p. 121.

Chapter 8: Séguin

1. Galton, "Hereditary Talent," p. 157.
2. Richardson, *Making of Intelligence*, p. 25.
3. Story of the Wild Boy is told by Itard in *The Wild Boy of Aveyron*, and by François Truffaut in his 1970 film *The Wild Child*.
4. Itard, *The Wild Boy of Aveyron*, pp. 55–66: "Development of the Functions of the Senses."
5. Scheerenberger, *A History of Mental Retardation*, p. 69.
6. The story of Samuel Howe and his famous student is told in two works: Freeberg, *The Education of Laura Bridgman*; Gitter, *Imprisoned Guest*.
7. Scheerenberger, *A History of Mental Retardation*, p. 121.
8. Remarks by Dr. George Brown from "In Memory of Édouard Séguin, M.D.," *Proceedings of the AMO*, 1880.
9. Rafter, *Creating Born Criminals*, p. 60.
10. Scheerenberger, *A History of Mental Retardation*, p. 68.
11. Ibid., p. 62.

12. Ibid.
13. Ibid., p. 68.
14. Ibid., p. 65.
15. Ibid., p. 157.
16. Galton, "Hereditary Talent," p. 166.
17. Scheerenberger, *A History of Mental Retardation*, p. 124.
18. Wilmarth, "Report on the Examination," p. 148.
19. Hall, "Laura Bridgman," p. 164.

Chapter 9: Guiteau

1. For a detailed account of the Guiteau trial, see Rosenberg, *Trial of the Assassin.*
2. Dana, "Early Neurology in the United States," p. 1422.
3. Wright's brain is described in Dwight, "Remarks on the Brain," pp. 210–15.
4. For details of the Kemmler fiasco, see Richard Moran, *Executioner's Current: Thomas Edison, George Westinghouse, and the Invention of the Electric Chair* (New York: Knopf, 2002).
5. For the state of physiology in the United States in the 1800s, see "S. Weir Mitchell, Philadelphia's Frustrated Physiologist and Triumphant Reformer," in Fye, *The Development of American Physiology*, pp. 54–91.
6. E. C. Spitzka, "Reform in the Scientific Study of Psychiatry," p. 702.
7. E. C. Spitzka, "Merits and Motives," p. 704.
8. Ibid., p. 706.
9. Blustein, "Hollow Square," p. 254.
10. Excerpts from Spitzka's testimony are contained in Rosenberg, *Trial of the Assassin*, pp. 155–77; and in Haines, "Spitzka and Spitzka," pp. 245–46.
11. Lamb's "Report on the Post-mortem Examination of the Body of Charles J. Guiteau" was published in *Medical News* 41, 8 July 1882, pp. 43–45; and 9 September 1882, pp. 297–99.
12. "Dahmer Family Split Over Use of Brain," *New York Times,* 3 August 1995, p. A22.
13. Mark Landler, "German Radical's Daughter Seeks Brain Kept After Suicide," *New York Times,* 11 November 2002, p. A8.
14. Blustein, "Hollow Square," p. 263.

Chapter 10: Wilder

1. Earnest, *S. Weir Mitchell*, p. 91.
2. See *The Autobiography of Isaac Jones Wistar* (Philadelphia: Wistar Institute of Anatomy and Biology, 1937).

3. The description is from S. Weir Mitchell's novel *Characteristics* (1915), pp. 134–35.
4. Davidson, *The Bone Sharp,* p. 109.
5. Warren, *Joseph Leidy, The Last Man Who Knew Everything.*
6. Ruloff's story is told in Kasprzak, p. 14.
7. Letter from Harrison Allen to Burt Green Wilder, 25 November 1882. Burt Green Wilder Papers (file 14/26/95), Kroch Library, Cornell University.
8. *Ithaca Daily Journal,* 15 February 1895.
9. Burt Green Wilder Papers (file 14/26/95), box 10.
10. *Ithaca Daily Journal,* 15 February 1895.
11. For the full text of the first bequest form, see Kasprzak, p. 5.
12. Wilder's letter of resignation, dated 20 December 1891, is in the Burt Green Wilder Papers (file 14/26/95), Kroch Library, Cornell University. Harrison Allen's response, dated 23 December 1891, is in the same file.
13. Ibid.
14. Wilder recounts the Wister episode in *Brain of the American Negro,* pp. 26–30.
15. Wilder's social activism is explored in Beardsley, "The American Scientist as Social Activist."
16. Wilder, *Brain of the American Negro,* pp. 1–2.
17. Burt G. Wilder, review of *America's Greatest Problem: The Negro,* by R. W. Schufeldt, *Science* n.s. 42, no. 1091, p. 768.
18. Letter from Edward A. Spitzka to Burt Green Wilder, 25 February 25, 1908. Burt Green Wilder Papers (file 14/26/95), Kroch Library, Cornell University.

Chapter 11: Whitman

1. For background on American phrenology, see John D. Davies, *Phrenology: Fad and Science* (New Haven: Yale University Press, 1955); and T. H. and G. E. Leahy, *Psychology's Occult Doubles* (Chicago: Nelson-Hall, 1983).
2. Emerson, *Conduct of Life,* "Fate."
3. For the phrenological charts of prominent Americans, see Madeleine B. Stern, ed., *A Phrenological Dictionary of Nineteenth-Century Americans* (Westport: Greenwood Press, 1982).
4. See Thomas L. Brasher, "Whitman's Conversion to Phrenology," *Walt Whitman Newsletter* 4 (June 1958), pp. 95–97.
5. Hungerford, "Walt Whitman and His Chart of Bumps," pp. 350–84; and Aspiz, *Walt Whitman*, pp. 3–33.
6. It would not be long before most phrenological readings would produce charts like Whitman's. Few paying customers received very many scores below 5 (there had to be a few areas that needed work). Whitman, for exam-

ple, was given low marks for humor, which caused him a good deal of puzzlement.

7. Traubel, *With Walt Whitman*, vol. 1, pp. 454–55.
8. See *Dictionary of American Biography*, vol. 13, p. 64.
9. *Gunshot Wounds* is included in S. Weir Mitchell et al., *Reflex Paralysis* (New Haven Historical Library: Yale University School of Medicine, 1941).
10. See Edmund Wilson, *The Wound and the Bow* (Boston: Houghton-Mifflin), p. 196.
11. Earnest, *S. Weir Mitchell*, p. 83.
12. Ibid.
13. Haymaker, *Founders of Neurology*, p. 482.
14. Traubel, *With Walt Whitman*, vol. 1, pp. 454–55.
15. Ibid., vol. 2, p. 271.
16. Earnest, *S. Weir Mitchell*, p. 233.
17. Cushing, *Life of Sir William Osler*, vol. 1, pp. 264–65.
18. For Holmes on phrenology, see Traubel, *With Walt Whitman*, vol. 1, p. 385.
19. Ibid., vol. 9, p. 598.
20. Ibid., p. 603.
21. Longaker, "The Last Sickness," p. 406.
22. Cattell, *Notes on the Demonstrations.*
23. "Autopsy Made on Whitman's Body," *Philadelphia Press*, 28 March 1892, p. 2.
24. Traubel, *With Walt Whitman*, vol. 9, p. 605.
25. Longaker, "The Last Sickness," p. 404. Traubel reprints one of Longaker's daily bulletins describing the same incident at greater length in *With Walt Whitman in Camden*, vol. 9, p. 587.
26. "Three Hundred Men Pledged Their Brains to Science," *New York Herald* 4 September 1898, section 5:6.

Chapter 12: Spitzka

1. Letter from Edward A. Spitzka to Joseph Leidy Jr., 25 April 25 1901. Philadelphia College of Physicians.
2. MacDonald and Spitzka, "The Trial," pp. 22–23.
3. Edward A. Spitzka, "Memo for the Use of the Members of the Committee of the American Anthropometric Society," 3 January 1906. Philadelphia College of Physicians.
4. "Brain Research by Phila. Anatomist Startles Science," *Philadelphia North American*, 5 December 1907, p. 1.
5. "Loss of Poet's Brain Roils His Executors," *Philadelphia North American*, 6 December 1907, p. 1.
6. E. A. Spitzka, "The Development of Man's Great Brain," p. 327.

7. E. A. Spitzka, "A Study of the Brains," p. 226.
8. E. C. Spitzka, quoted in E. A. Spitzka, "Study of the Brain of the Late Major J. W. Powell," p. 603.
9. E. A. Spitzka, "The Development of Man's Great Brain," p. 325.
10. "Scientists Attack Dr. Spitzka's Brain Structure Theory," *Philadelphia North American,* 6 December 1907, p. 1.
11. Bean, "The Negro Brain," p. 784.
12. "To Have Us Know More About Brains," *New York Times,* 9 April 1911, p. 11.
13. Mall, "On Several Anatomical Characters," p. 7.
14. Bean, "The Negro Brain," p. 784.
15. Mall, "On Several Anatomical Characters," p. 9.
16. Ibid.
17. Clause, "The Wistar Rat," pp. 329–49.
18. Kennedy, *Fight of a Book for the World,* pp. 139–40.
19. Donaldson and Canavan, "A Study of the Brains," p. 83.
20. "Brain of Executed Slayer [Gillette] Was Normal, Says Phila. Expert," *Philadelphia North American,* 1 April 1908.
21. "Dr. Spitzka, Brain Expert, in Nervous Breakdown," *Philadelphia Press,* 22 March 1913, p. 1.
22. E. C. Spitzka's conspiracy theory was spelled out in "The Workings of Anarchists, Understandings That Are International with a Directing Mind in London," *New York Times,* 6 June 1909, p. 4.
23. "Husband Sues Dr. Spitzka," *New York Times,* 13 November 1913, p. 1.
24. "Dr. E. C. Spitzka, Alienist, Is Dead," *New York Times,* 14 January 1914, p. 3.
25. Information about Eakins's Spitzka portrait was provided by Douglas Paschall of the Philadelphia Museum of Art and Phylis Rosenzweig of the Hirshhorn Museum.
26. Feindel, "Osler's Brain Again," p. 1.

Chapter 13: Lenin

1. Irina Bogolepova—statement of research interest, 18 September 2000.
2. Haymaker and Schiller, *Founders of Neurology,* p. 7.
3. Campbell's map was first published in Alfred Walter Campbell, *Histological Studies on the Localization of the Cerebral Cortex* (Cambridge: University Press, 1905); for Elliot Smith's cortical map, see *Journal of Anatomy and Physiology* 41 (1907), pp. 237–54; for Brodmann's map, see Korbinian Brodmann, *Vergleichende Lokalisationslehre der Grosshirnrinde in ihren Prinzipien dargestellt auf Grund des Zellenbaues* (Leipzig: Barth, 1909).
4. The activities of the Brain Commission are described in Jochen Richter, "The

Brain Commission of the International Association of Academies: the First International Society of Neurosciences," unpublished manuscript.

5. Wilson, *My Memoir*, p. 289.
6. Sarkisov, *Structure and Functions of the Brain*, p. 239.
7. An irreverent and sensationalized account of Vogt's life can be found in Spengler, *Lenin's Brain*; a much more sentimental one in Klatzo, *Cécile and Oskar Vogt*.
8. For an in-depth look at Vogt's accomplishments as a hypnotist, see Alan Gould, *A History of Hypnotism* (Cambridge University Press, 1992), pp. 421–22, 537–43.
9. "The Brain of Lenin," *Journal of the American Medical Association*, 3 March 1928, p. 708.
10. Klatzo, *Cécile and Oskar Vogt*, p. 118.
11. Oskar Vogt, "Bericht über die Arbeiten," p. 109.
12. Constantin Freiherr von Economo and George N. Koskinas, *Die Cytoarchitektonic der Hirnrinde des erwachsenen Menschen* (Vienna and Berlin: Springer, 1925).
13. Bentivoglio, "Cortical Structure," p. 294.
14. Oskar Vogt, "Bericht über die Arbeiten," p. 111.
15. Wilder Penfield, "Going-off on a Tangent: Wilder Penfield's Memoirs of His Visit with the Vogt, in Berlin, in 1928," reprinted in Kreutzberg et al., "Oskar and Cécile Vogt," p. 370.
16. Constantin von Economo, "Problems of Brain Research," address given in Vienna, 7 May 1931.
17. Oskar Vogt, "Bericht über die Arbeiten," p. 113.
18. Tucker, *Stalin in Power*, p. 553.
19. Vogt's Black Forest saga is described in Kreutzberg et al., "Oskar and Cécile Vogt," pp. 363–71.
20. See, for example, S. Blinkov and G. Poliakov, "The Activities of the Moscow Brain Institute," pp. 674–79.
21. Adrianov, "Nerasgadanni Tainii," p. 3.
22. Bailey and von Bonin, *Isocortex of Man*, p. 59.
23. Ibid., pp. 232–33.

Chapter 14: Einstein

1. An account of Einstein's autopsy and the removal of his eyeballs is given in Brian, pp. 437–39.
2. The saga of Harvey's role in the fate of Einstein's brain is told in Maranto, "Einstein's Brain," pp. 28–34.
3. Kantha, "Albert Einstein's Dyslexia," pp. 119–22.

4. Hines, "Further on Einstein's Brain," pp. 343–44.
5. Anderson and Harvey, "Alterations in Cortical Thickness," pp. 161–64.
6. D. L. Kigar et al., "Estimates of Cell Number in Temporal Neocortex," pp. 89, 213.
7. Retzius, "Das Gehirn des Astronomen Hugo Gyldén," pp. 1–22; "Das Gehirn der Mathematikerin Sonja Kovalevsky," pp. 1–16.
8. For a discussion of criminal brains and confluent fissures, see C. W. M. Poynter, "A Study of Cerebral Anthropology, with a Description of Two Brains of Criminals," *University Studies* 12 no. 4 (October 1912), pp. 345–439.
9. Anderson and Harvey, "Alterations in Cortical Thickness," p. 161.
10. For a good discussion of the lack of consensus on intelligence, see Richardson, *Making of Intelligence.*
11. *Bulletin de la Société d'Anthropologie,* 2 May 1861, p. 311.
12. Jerry Fodor, "Making the Connection," *Times Literary Supplement,* 17 May 2002, p. 3.
13. Uttal, *New Phrenology,* p. 184.
14. "If You're Not Musical Now You Never Will Be," *New Scientist,* 22 June 2002, p. 25.
15. For studies of musicians' brains, see Auerbach, "Beitrag zur Lokalisation"; and Meyer, "Search for a Morphological Substrate."

Chapter 15: The New Phrenologists

1. Stephen Pinker, "His Brain Measured Up," *New York Times,* 24 June 1999, p. A27.
2. E. A. Spitzka, "A Study of the Brains," p. 233.
3. Gould, *Mismeasure of Man,* p. 127.
4. Interview with the author, December 2001.
5. Francis Bacon, *The Advancement of Learning* (1605), bk. I, v. 8.
6. See Stott, *Darwin and the Barnacle.*
7. Gould, *Mismeasure of Man,* p. 22.

Bibliography

Ackerknecht, Erwin H., and Henri V. Vallois. *Franz Josef Gall, Inventor of Phrenology and His Collection.* Translated by Claire St. Léon. Madison: University of Wisconsin Medical School, 1956.

Adrianov, Oleg S. "Issledovanie mozga Lenina." *Ospekhi fiziologicheskikh nauk* 24, no. 3 (1993): 40–52.

———. "Nerasgadanni Tainii [Unsolved Mysteries]: Interview with Oleg Adrianov." *Nauka i Religia* [Science and Religion] (8 August 1989): 2–4.

Amunts, V. V. "Individual Cytoarchitectonic Peculiarities of Human Subcortical-Brain Stem Cerebral Structures." *Zhurnal nevropatologii i psykhiatrii imeni S.S. Korsakova* 97, no. 3 (1997): 49–52.

Anderson, Britt, and Thomas Harvey. "Alterations in Cortical Thickness and Neuronal Density in the Frontal Cortex of Albert Einstein." *Neuroscience Letters* 210 (1996): 161–64.

Aspiz, Harold. *Walt Whitman and the Body Beautiful.* Urbana: University of Illinois Press, 1980.

Auerbach, Sigmund. "Beitrag zur Lokalisation des musikalischen Talentes im Gehirn und am Schädel." *Archiv für Anatomie und Physiologie* (1906): 197–230.

Babbage, Neville F. "Autopsy Report on the Body of Charles Babbage ('Father of the Computer')." *Medical Journal of Australia* 154 (1991): 758–59.

Babcock, Warren L. "On the Morbid Heredity and Predisposition to Insanity of the Man of Genius." *Journal of Nervous and Mental Disease* 20, no. 12 (December 1895): 749–69.

BIBLIOGRAPHY

Bailey, Percival. "The Seat of the Soul." *Perspectives in Biology and Medicine* 2 (Summer 1959): 417–40.

Bailey, Percival, and Gerhardt von Bonin. *The Isocortex of Man.* Urbana: University of Illinois Press, 1951.

Bastian, H. Charlton. *The Brain as the Organ of the Mind.* New York: D. Appleton, 1880.

Bean, Robert Bennett. "The Negro Brain." *The Century Magazine* 76 (October 1906): 778–84.

———. "The Training of the Negro." *The Century Magazine* 72 (September 1906): 947–53.

Beardsley, Edward H. "The American Scientist as Social Activist: Franz Boas, Burt G. Wilder, and the Cause of Racial Justice, 1900–1915." *Isis* 64, no. 221 (March 1973): 50–66.

Beccaria, Cesare. *Of Crimes and Punishments.* Translated by Jane Grigson. New York: Marsilio, 1996.

Benedikt, Moritz. *Anatomical Studies upon the Brains of Criminals.* Translated by E. P. Fowler. New York: W. Wood, 1881.

Bentivoglio, Marina. "Cortical Structure and Mental Skills: Oskar Vogt and the Legacy of Lenin's Brain." *Brain Research Bulletin* 47, no. 4 (1998): 291–96.

Berlin, Isaiah. "Historical Inevitability." *Four Essays on Liberty.* London: Oxford, 1969.

Blinkov, S., and G. Poliakov. "The Activities of the Moscow Brain Institute." *Acta Medica URSS,* 1, no. 3–4 (1938): 674–79.

Blustein, Bonnie Ellen. " 'A Hollow Square of Psychological Science': American Neurologists and Psychiatrists in Conflict." In *Madhouses, Mad-Doctors, and Madmen: The Social History of Psychiatry in the Victorian Era.* Edited by Andrew Scull. Philadelphia: University of Pennsylvania Press, 1981.

Bogolepova, I. N., and N. N. Bogolepov. "The Brain of V. V. Mayakovsky." *Zhurnal Nevropatologi psykhiatrii* 97, no. 5 (1997): 47–50.

Bower, James M., and Lawrence M. Parsons. "Rethinking the 'Lesser Brain.' " *Scientific American* (August 2003): 51–57.

"The Brains of Nikolai Lenin and Anatole France." *Science* 66, no. 1721 (1927): xii.

Brian, Denis. "Einstein's Brain." Appendix in *Einstein: A Life.* New York: Wiley, 1996.

Bühler, W. K. *Gauss: A Biographical Study.* Berlin: Springer-Verlag, 1981.

Burrell, Brian. "The Strange Fate of Whitman's Brain." *Walt Whitman Quarterly Review* 20 (Winter/Spring 2003): 107–33.

Cattell, Henry W. *Notes on the Demonstrations in Morbid Anatomy (Including Autopsies) Delivered in the Medical Department of the University of Pennsylvania Before the Third-Year Class.* Philadelphia: International Medical Magazine Co., 1894.

Clarke, Edwin, and L. S. Jacyna. *Nineteenth-Century Origins of Neuroscientific Concepts.* Berkeley: University of California Press, 1987.

Clarke, Edwin, and C. D. O'Malley. *The Human Brain and Spinal Cord: A Historical Study Illustrated by Writings from Antiquity to the Twentieth Century.* Berkeley: University of California Press, 1968.

Clause, Bonnie Tocher. "The Wistar Rat as a Right Choice: Establishing Mammalian Standards and the Ideal of a Standardized Mammal." *Journal of the History of Biology* 26, no. 2 (Summer 1993): 329–49.

Colombo, Giorgio. *La scienza infelice: Il museo di antropologia criminale di Cesare Lombroso.* Turin: Boringhieri, 1975.

Comte, Auguste. *Auguste Comte and Positivism: The Essential Writings.* Edited by Gertrude Lenzer. New York: Harper and Row, 1975.

Cushing, Harvey. *The Life of Sir William Osler.* Volume 1. Oxford: Clarendon Press, 1925.

Cutler, Allan. *The Seashell on the Mountaintop: A Story of Science, Sainthood, and the Humble Genius Who Discovered a New History of the Earth.* New York: Dutton, 2003.

Dana, Charles L. "Early Neurology in the United States." *Journal of the American Medical Association* 90 no. 18 (5 May 1928): 1421–24.

Davidson, Jane Pierce. *The Bone Sharp.* Philadelphia: Academy of Natural Sciences, 1997.

Descartes, René. *The Philosophical Works of Descartes.* Translated by Elizabeth S. Haldane and G. R. T. Ross. New York: Dover, 1955.

———. *Principles of Philosophy.* Translated by Valentine Rodger Miller and Reese P. Miller. Dordrecht: Reidel, 1983.

Diamond, Marian et al. "On the Brain of a Scientist: Albert Einstein." *Experimental Neurology* 88 (1985): 198–204.

Donaldson, Henry H. "The Significance of Brain Weight." *Archives of Neurology and Psychiatry* 13 (March 1925): 385–86.

———. "Anatomical Observations on the Brain and Several Sense-Organs of the

Blind Deaf-Mute, Laura Dewey Bridgman." *American Journal of Psychology* 3 (September 1890): 293–342; and 4 (December 1891): 248–94.

———. "The Extent of the Visual Area of the Cortex in Man, as Deduced from the Study of Laura Bridgman's Brain." *American Journal of Psychology* 4 (August 1892): 503–13.

Donaldson, Henry H., and Myrtelle M. Canavan, "A Study of the Brains of Three Scholars: Granville Stanley Hall, Sir William Osler, and Edward Sylvester Morse." *Journal of Comparative Neurology* 46 (15 August 1928): 1–95.

Dunnington, Guy Waldo. *Carl Friedrich Gauss, Titan of Science: A Study of His Life and Work.* New York: Exposition Press, 1955.

Dwight, Thomas. "Remarks on the Brain, Illustrated by the Description of the Brain of a Distinguished Man [Chauncey Wright]." *Proceedings of the American Academy of Arts and Sciences* 13 (1877–1878): 210–15.

Earnest, Ernest P. *S. Weir Mitchell, Novelist and Physician.* Philadelphia: University of Pennsylvania Press, 1950.

Economo, Constantin von. "Some New Methods for Studying Brains of Exceptional People (Encephalometry and Brain Casts)." *Journal of Nervous and Mental Diseases* 72, no. 2 (August 1930): 300–302.

———. "Wie sollen wir Elitegehirne verarbeiten?" *Zeitschrift für Psychologie & Physiologie* 121 (1929): 323–409.

Feindel, William. "Osler's Brain Again." *Osler Library Newsletter* 64 (June 1990): 1.

Finger, Stanley. "Intellect and the Brain." In *Origins of Neuroscience: A History of Explorations into Brain Function.* New York: Oxford University Press, 1994.

———. *Minds Behind the Brain: A History of the Pioneers and Their Discoveries.* London: Oxford University Press, 2000.

Freeberg, Ernest. *The Education of Laura Bridgman: First Deaf and Blind Person to Learn Language.* Cambridge: Harvard University Press, 2001.

Fye, W. Bruce. *The Development of American Physiology: Scientific Medicine in the Nineteenth Century.* Baltimore: Johns Hopkins University Press, 1987.

Galton, Francis. *Hereditary Genius: An Inquiry into Its Laws and Consequences.* New York: D. Appleton, 1879.

———. "Hereditary Talent and Character." *Macmillan's Magazine* 12 (1865): 156–166, 318–27.

Gibson, Mary. *Born to Crime: Cesare Lombroso and the Origins of Biological Criminology.* Westport: Praeger, 2002.

Gitter, Elisabeth. *The Imprisoned Guest: Samuel Howe and Laura Bridgman, the Original Deaf, Blind Girl.* New York: Farrar, Straus and Giroux, 2001.

Gould, Stephen Jay. *The Mismeasure of Man.* Revised edition. New York: W. W. Norton, 1996.

———. *Ontogeny and Phylogeny.* Cambridge, Mass.: Belknap Press, 1977.

Graham, Loren R. *Science and Philosophy in the Soviet Union.* New York: Knopf, 1972.

Gratiolet, Pierre-Louis. *Mémoire sur les plis cérébraux de l'homme et des primates.* Paris: A. Bertrand, 1854.

Gregory, Frederick. *Scientific Materialism in Nineteenth-Century Germany.* Boston: D. Reidel, 1977.

Hagner, Michael. "Cultivating the Cortex in German Neuroanatomy." *Science in Context* 14 no. 4 (2001): 541–63.

———. "Skulls, Brains, and Memorial Culture: On Cerebral Biographies of Scientists in the Nineteenth Century." *Science in Context* 16 no. 1–2 (2003): 195–218.

Haines, Duane E. "Spitzka and Spitzka on the Brains of Assassins." *Journal of the History of the Neurosciences* 4, no. 3–4 (1995): 236–66.

Hall, G. Stanley. "Laura Bridgman." *Mind: A Quarterly Review of Psychology and Philosophy* 14 (April 1879): 149–72.

Hammond, Michael. "Anthropology as a Weapon of Social Combat in Late-Nineteenth-Century France." *Journal of the History of the Behavioral Sciences* 16 (1980): 118–32.

Hansemann, David. "Ueber das Gehirn von Hermann v. Helmholtz." *Zeitschrift fur Psychologie & Physiologie* 20 (1899): 1–12.

———. *Ueber die Gehirne von Th. Mommsen, Historiker, R. W. Bunsen, Chemiker, und Ad. v. Menzel, Maler.* Stuttgart: E. Schweizerbart, 1907.

Hartley, Lucy. *Physiognomy and the Meaning of Expression in Nineteenth-Century Culture.* Cambridge, Eng.: Cambridge University Press, 2000.

Haymaker, Webb, and Francis Schiller, eds. *The Founders of Neurology.* 2nd edition. Springfield, Ill.: Thomas, 1970.

Hecht, Jennifer Michael. *The End of the Soul: Scientific Modernity, Atheism, and Anthropology in France.* New York: Columbia University Press, 2003.

Henschen, Salomon E. *Klinische und anatomische Beiträge zur Pathologie des Gehirns.* Uppsala: Almquist und Wiksells, 1890–1903.

Hines, Terence. "Further on Einstein's Brain." *Experimental Neurology* 150 (1998): 343–44.

Horsley, Victor. "Description of the Brain of Mr. Charles Babbage, F.R.S." *Philosophical Transactions of the Royal Society*, series B, 200 (1908): 117–31.

Humphry, G. M. "Case of Injury to the Brain. (Being an Account of the Case of the Late Rev. Dr. Whewell, Master of Trinity College, Cambridge.)" *Lancet* 17 (March 1866): 279–80.

Hungerford, Edward. "Walt Whitman and His Chart of Bumps." *American Literature* 2 (January 1931): 350–84.

Iserson, Kenneth V. *Death to Dust: What Happens to Dead Bodies?* Tucson: Galen Press, 1994.

Itard, Jean Marc Gaspard. *The Wild Boy of Aveyron.* Translated by George and Muriel Humphrey. New York: Appleton, 1962.

Jacobsen, Marcus. *Foundations of Neuroscience.* New York: Plenum Press, 1993.

Jeannerod, Marc. *The Brain Machine.* Cambridge, Mass.: Harvard University Press, 1985.

Kantha, S. S. "Albert Einstein's Dyslexia and the Significance of Brodmann Area 39 of His Left Cerebral Cortex." *Medical Hypotheses* 37 (1992): 119–22.

Kasprzak, Hedwig. "Report on the Wilder Brain Collection" (pamphlet). Ithaca: Cornell University, 1972.

Kennedy, William Sloane. *Fight of a Book for the World.* West Yarmouth: Stonecraft Press, 1926.

Kigar, D. L., Sandra Witelson, I. I. Glezer, and Thomas Harvey. "Estimates of Cell Number in Temporal Neocortex in the Brain of Albert Einstein." *Society for Neuroscience Abstracts* 23, no. 89.9 (1997).

Klatzo, Igor. *Cécile and Oskar Vogt: The Visionaries of Modern Neuroscience.* Vienna; New York: Springer-Verlag, 2002.

Kretschmer, Ernst. *The Psychology of Men of Genius.* Translated by R. B. Cattell. New York: Harcourt, Brace and Co., 1931.

Kreutzberg, Georg W., Igor Klatzo, and Paul Kleihues. "Oskar and Cécile Vogt, Lenin's Brain and the Bumble-Bees of the Black Forest." *Brain Pathology* 2 (1992): 363–71.

Kuhn, Thomas S. *The Structure of Scientific Revolutions.* Chicago: University of Chicago Press, 1962.

La Mettrie, Julien Offray de. *Man a Machine.* Edited by G. C. Bussy. LaSalle: Open Court, 1912.

Lashley, K. S., and George Clark. "The Cytoarchitecture of the Cerebral Cortex of Ateles: A Critical Examination of Architectonic Studies." *Journal of Comparative Neurology* 85, no. 2 (1946): 223–305.

Leahy, T. H. *Psychology's Occult Doubles: Psychology and the Problem of Pseudoscience.* Chicago: Nelson-Hall, 1983.

Leon, Philip W. *Walt Whitman and Sir William Osler: A Poet and His Physician.* Toronto: ECW Press, 1995.

Leuret, François. *Du Traitement Moral de la Folie.* Paris: J. B. Baillière, 1840.

Lombroso, Cesare. *The Man of Genius.* London: Walter Scott, 1905.

Longaker, Daniel. "The Last Sickness and Death of Walt Whitman." In *In Re Walt Whitman.* Edited by Horace Traubel et al. Philadelphia: David McKay, 1893.

MacDonald, Carlos F., and Edward Anthony Spitzka. "The Trial, Execution, Autopsy, and Mental Status of Leon F. Czolgosz, alias Fred Nieman, Assassin of President McKinley, with a Report of the Post-Mortem Examination." *New York Medical Journal* (4 January 1902): 12–23.

Macmillan, Malcolm. *An Odd Kind of Fame: Stories of Phineas Gage.* Cambridge, Mass.: MIT Press, 2000.

Mall, Franklin P. "On Several Anatomical Characters of the Human Brain, Said to Vary According to Race and Sex, with Especial Reference to the Weight of the Frontal Lobe." *American Journal of Anatomy* 9, no. 1 (1909): 1–32.

Maranto, Gina. "Einstein's Brain." *Discover* (May 1985): 28–34.

Marchand, Leslie A. *Byron: A Biography.* New York: Knopf, 1957.

Marshall, John. "On the Relations Between the Weight of the Brain and Its Parts, and the Stature and Mass of the Body in Man." *Journal of Anatomy and Physiology* 26 (July 1892): 445–500.

———. "On the Brain of the Late George Grote, F. R. S., with Comments and Observations on the Human Brain and Its Parts Generally." *Journal of Anatomy and Physiology, Normal and Pathological* 27 (October 1892): 21–65.

Marshall, Louise H., and Horace W. Magoun. *Discoveries in the Human Brain.* Totowa: Humana Press, 1998.

Martensen, Robert. "When the Brain Came Out of the Skull: Thomas Willis (1621–1675), Anatomical Technique and the Formation of the 'Cerebral Body' in Seventeenth-Century England." In *A Short History of Neurology.* Edited by F. Clifford Rose. Oxford: Butterworth-Heinemann, 1999.

Meyer, Alfred. "The Search for a Morphological Substrate in the Brains of Eminent Persons Including Musicians: A Historical Review." In *Music and the Brain.*

Edited by MacDonald Critchley and R. A. Henson. London: William Heinemann, 1977.

Mill, John Stuart. *The Correspondence of John Stuart Mill and Auguste Comte.* Translated by Oscar A. Haac. New Brunswick: Transaction, 1995.

Mountcastle, Vernon. *Perceptual Neuroscience: The Cerebral Cortex.* Cambridge, Mass.: Harvard University Press, 1998.

Nagayo, Mataro, Yushi Uchimara, and Shiho Nishimaru. *Kesshutsu jinno no kenkyu* [Studies of Brains of Distinguished People]. Tokyo: Iwanami Shoten, 1939.

Naumov, Anatolii V. *Rossiiskoe Ugolovnoe Pravo, Oshchaia chast.* Moscow: Beck, 1997.

Oeser, E., and F. Seitelberger. "From Brain Pathology to Neurophilosophy." In *Neuroscience Across the Centuries.* Edited by F. Clifford Rose. London: Smith-Gordon, 1989.

Osler, William. "The Leaven of Science." In *Aequanimitas.* 2nd edition. Philadelphia: P. Blakiston's Son and Co., 1919.

———. "On the Brains of Criminals, with a Description of the Brains of Two Murderers." *Canada Medical & Surgical Journal* 10 (1882): 385–98.

Papez, James W. "The Brain of Burt Green Wilder." *Journal of Comparative Neurology* 47 (1929): 285–322.

———. "The Brain of Sutherland Simpson." *Journal of Comparative Neurology* 51 (1930): 165–96.

Parker, Andrew J. "Morphology of the Cerebral Convolutions with Special Reference to the Order of Primates." *Journal of the Academy of Natural Sciences* 10, 2nd series (1894–1896): 247–365.

Paterniti, Michael. *Driving Mr. Albert.* New York: Dial Press, 2000.

Pearl, R. "Biometrical Studies on Man: I. Variation and Correlation in Brain Weight." *Biometrika* 4 (June 1905): 13–104.

———. "On the Correlation Between Intelligence and the Size of the Head." *Journal of Comparative Neurology and Psychology* 16 (1906): 189–99.

Pearson, Karl. "On the Correlation of Intellectual Ability with the Size and Shape of the Head." *Proceedings of the Royal Society of London* 69 (1901–1902): 333–42.

Pick, Daniel. *Faces of Degeneration.* Cambridge, Eng.: Cambridge University Press, 1989.

Pinker, Steven. "His Brain Measured Up." *New York Times,* 24 June 1999, p. A27.

The Podromus of Nicolaus Steno's Dissertation. Translated by John Garrett Winter. London: Macmillan, 1916.

Possony, Stefan. *Lenin: The Compulsive Revolutionary*. Chicago: Regnery, 1964.

Poynter, C. W. M. "A Study of Cerebral Anthropology, with a Description of Two Brains of Criminals." *University Studies* 12:4 (October 1912): 345–439.

Quatrefages, Armand de. *Rapport sur les Progrès de l'Anthropologie*. Paris: Imprime Impériale, 1867.

Quetelet, Adolphe. *A Treatise on Man and the Development of His Faculties*. Reprint of the 1842 English translation. Gainsville, Fla.: Scholars' Facsimiles and Reprints, 1969.

Rafter, Nicole Hahn. *Creating Born Criminals*. Urbana and Chicago: University of Illinois Press, 1997.

Reise, Walter. "The Brain of Trigant Burrow, Physician, Scientist, and Author." *Journal of Comparative Neurology* 100, no. 3 (June 1954): 525–67.

———. "Brains of Prominent People: History, Facts and Significance." *Medical College of Virginia Quarterly* 2 (1966): 106–10.

Reise, Walter, and Kurt Goldstein. "The Brain of Ludwig Edinger: An Inquiry into the Cerebral Morphology of Mental Ability and Left-Handedness." *Journal of Comparative Neurology* 92, no. 2 (April 1950): 133–61.

Retzius, Gustav. "Das Gehirn des Astronomen Hugo Gyldén." *Biologische Untersuchungen* 8 (1898): 1–22; "Das Gehirn der Mathematikerin Sonja Kovalevsky." *Biologische Untersuchungen*. 9 (1900): 1–16.

Richardson, Ken. *The Making of Intelligence*. New York: Columbia University Press, 2000.

Richter, Jochen. "Zytoarchitektonik und Revolution—Lenins Gehirn als Raum und Objekt." *Berichte zur Wissenschaftsgeschichte* 23 (2000): 1–16.

Roberts, Colin M. "Edward A. Spitzka: The Anthropometric Society and the Struggle to Map Great Brains." In *Jefferson Medical College: Legend and Lore*. Edited by F. B. Wagner and J. W. Savacool. Philadelphia: William T. Cooke, 1996.

Rodin, Alvin, and Jack D. Key. "Osler's Brain and Related Mental Matters." *Southern Medical Journal* 83, no. 2 (February 1990): 207–12.

Rosenberg, Charles E. *The Trial of the Assassin Guiteau*. Chicago: University of Chicago Press, 1968.

Sagan, Carl. *Broca's Brain: Reflections on the Romance of Science*. New York: Random House, 1979.

Sarkisov, S. A. *The Structure and Functions of the Brain.* Translated by Basil Haigh. Bloomington: Indiana University Press, 1966.

Schiller, Francis. *Paul Broca: Founder of French Anthropology, Explorer of the Brain.* Berkeley: University of California Press, 1979.

Schneerenberger, R. C. *A History of Mental Retardation.* Baltimore: Paul H. Brooks, 1983.

Schneider, Peter et al. "Morphology of Heschl's Gyrus Reflects Enhanced Activation in the Auditory Cortex of Musicians." *Nature Neuroscience* 5 (2002): 688–94.

Séguin, Édouard. *Traitement moral et hygiène et éducation des idiots.* Paris: Association pour l'étude de l'histoire de la sécurité sociale, 1846.

Sekula, Allan. "The Body and the Archive." *October* 39 (1986): 3–64.

Shelley, Mary. *Frankenstein: The Modern Prometheus.* London: Wm. Pickering, 1996.

Simms, Joseph. "Human Brain-Weights." *Popular Science Monthly* 31 (1887): 355–59.

Spengler, Tilman. *Lenin's Brain.* Translated by Shaun Whiteside. New York: Farrar, Straus and Giroux, 1993.

Spitzka, Edward Anthony. "The Development of Man's Great Brain." *Connecticut Magazine* 9, no. 2 (1905): 319–29.

———. "Observations Regarding the Infliction of the Death Penalty by Electricity." *Proceedings of the American Philosophical Society* 47 (1908): 39–50.

———. "A Preliminary Communication of a Study of the Brains of Two Distinguished Physicians, Father and Son." *Philadelphia Medical Journal* 7 (6 April 1901): 680–88.

———. "A Study of the Brain of the Late Major J. W. Powell." *American Anthropologist* 5 (1903): 585–643.

———. "A Study of the Brain-Weights of Men Notable in the Professions, Arts and Sciences." *Philadelphia Medical Journal* 11 (May 1903): 757–61.

———. "A Study of the Brains of Six Eminent Scientists and Scholars Belonging to the American Anthropometric Society, Together with a Description of the Skull of Professor E. D. Cope." *Transactions of the American Philosophical Society,* n.s., 21 (1907): 175–308.

Spitzka, Edward Anthony, and H. E. Radasch. "The Brain Lesions Produced by Electricity as Observed After Legal Execution." *American Journal of the Medical Sciences* 144 (1912): 341–47.

Spitzka, Edward Charles. *Insanity: Its Classification, Diagnosis, and Treatment.* New York: Bermingham and Co., 1883.

———. "Merits and Motives of the Movement for Asylum Reform." *Journal of Nervous and Mental Disease* 5 (1878): 694–714.

———. "Reform in the Scientific Study of Psychiatry." *Journal of Nervous and Mental Disease* 5 (1878): 201–28.

Spivak, Monika. *Pasmertnaya Diagnostika Genialnostii [The Postmortem Diagnosis of Genius]: Edouard Bagritski, Andre Bely, Vladimir Mayakovski.* Moscow: Agraf, 2001.

Steno, Nicolaus. "Discours sur l'anatomie du cerveau." In *Leven en werk van Niels Stensen (1638–1686).* Edited by J. G. Vugs. Leiden: Universitaire Pers, 1968.

Stott, Rebecca. *Darwin and the Barnacle: The Story of One Tiny Creature and History's Most Spectacular Scientific Breakthrough.* New York: W. W. Norton, 2003.

Tobias, Phillip V. "Brain-Size, Grey Matter and Race—Fact or Fiction?" *American Journal of Physical Anthropology,* n.s., 32 (1970): 3–26.

Todd, T. Wingate. "A Liter and a Half of Brains." *Science* 66, no. 1701 (5 August 1927): 122–25.

Traubel, Horace. *With Walt Whitman in Camden,* 9 vols. Various publishers, 1906–1996.

Treffert, Darold A., and Gregory L. Wallace. "Islands of Genius." *Scientific American* (June 2002): 76–85.

Tucker, Robert C. *Stalin in Power: The Revolution from Above, 1928–1941.* New York: W. W. Norton, 1990.

Uttal, William R. *The New Phrenology: The Limits of Localizing Cognitive Processes in the Brain.* Cambridge, Mass.: MIT Press, 2001.

Vogt, Carl. *Köhlerglaube und Wissenschaft: Ein Streitschrift gegen Hofrath Rudolf Wagner in Göttingen.* Giessen: J. Rickersche, 1856.

———. *Lectures on Man, His Place in Creation, and in the History of the Earth.* Edited by James Hunt. London: Longman, 1864.

Vogt, Oskar. "Bericht über die Arbeiten des Moskauer Staatsinstituts für Hirnforschung." *Journal f. Psychologie und Neurologie* 40 (1929): 108–18.

Volkogonov, Dmitri. *Lenin: A New Biography.* Translated and edited by Harold Shukman. New York: Free Press, 1994.

Wagner, Rudolf. *Gespräche mit Carl Friedrich Gauss in den letzten Monaten seines Lebens.* Göttingen: Vandenhoeck and Ruprecht, 1975.

———. *Menschenschöpfung und Seelensubstanz.* Göttingen: Georg H. Wigand, 1854.

———. *Vorstudien zu einer wissenschaftlichen Morphologie und Physiologie des menschlichen Gehirns als Seelenorgan.* Göttingen: Verlag der Dieterichschen, 1860 and 1862.

Warren, Leonard. *Joseph Leidy: The Last Man Who Knew Everything.* New Haven, Conn.: Yale University Press, 1998.

Wilder, Burt Green. *The Brain of the American Negro.* New York: National Negro Committee, 1909.

———. "Brain: Methods of Removing, Preserving, Dissecting, and Drawing." *Reference Handbook of the Medical Sciences.* Edited by Albert H. Buck. New York: William Wood, 1900–1908.

Williams, Peter L. et al. *Gray's Anatomy.* 37th edition. Edinburgh: Churchill Livingstone, 1989.

Wilmarth, A. W. "Report on the Examination of One Hundred Brains of Feeble-Minded Children." *Proceedings of the Association of Medical Officers of American Institutions for Idiotic and Feeble-Minded Children* (1890): 138–48.

Wilson, Edith Bolling. *My Memoir.* Indianapolis: Bobbs-Merrill, 1939.

Witelson, Sandra, Debra L. Kigar, and Thomas Harvey. "The Exceptional Brain of Albert Einstein." *Lancet* 353 (19 June 1999): 2149–53.

Wittman, Axel D., Jens Frahm, and Wolfgang Hänicke. "Magnetresonanz-Tomografie des Gehirns von Carl Friedrich Gauss." *Gauss-Gesellschaft E.V. Göttingen Mitteilingen,* no. 36 (1999).

Wolfgang, Marvin E. "Cesare Lombroso." In *Pioneers in Criminology.* Edited by Hermann Mannheim. London: Stevens and Sons, 1960.

Young, Robert M. *Mind, Brain and Adaptation in the Nineteenth Century: Cerebral Localization and Its Biological Context from Gall to Ferrier.* Oxford: Clarendon Press, 1970.

Index

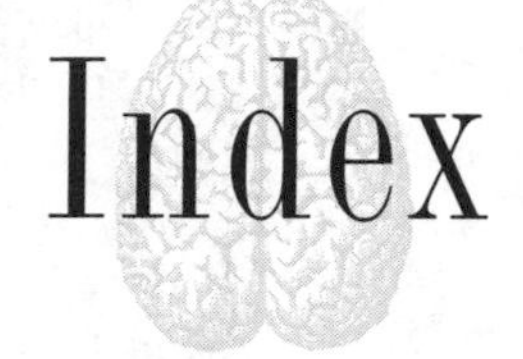

INDEX